VIBROACTIVITY OF BRANCHED AND RING STRUCTURED MECHANICAL DRIVES

APPLICATIONS OF VIBRATION SERIES

K. Ragulskis, Editor

K. Ragulskis, R. Bansevičius, R. Barauskas, and G. Kulvietis, Vibromotors for Precision Microrobots

P. Alabuzhev, A. Gritchin, L. Kim, G. Migirenko, V. Chon, and P. Stepanov, Vibration Protecting and Measuring Systems with Quasi-Zero Stiffness

S. Korablev, V. Shapin, and Yu. Filatov, Vibration Diagnostics in Precision Instruments

I. Vulfson, Vibroactivity of Branched and Ring Structured Mechanical Drives

IN PREPARATION

K. Ragulskis and Yu. Yurkauskas, Vibration of Bearings

VIBROACTIVITY OF BRANCHED AND RING STRUCTURED MECHANICAL DRIVES

I. Vulfson
Leningrad Institute of Textile and Light Industry

English Edition Editor
E. Rivin
Department of Mechanical Engineering
Wayne State University
Detroit, Michigan

HEMISPHERE PUBLISHING CORPORATION
A member of the Taylor & Francis Group
New York Washington Philadelphia London

VIBROACTIVITY OF BRANCHED AND RING STRUCTURED MECHANICAL DRIVES

Translated by V. Talačka.
Originally published as *Vibroaktivnost' privodov mashin razvetvlennoy i kol'tsevoy struktury* by Mashinostroyeniye, Leningrad, 1986.

1 2 3 4 5 6 7 8 9 0 BRBR 8 9 8 7 6 5 4 3 2 1 0 9

This book was set in Times Roman by Hemisphere Publishing Corporation. The production supervisor was Miriam Gonzalez and the typesetter was Vera Kishinevsky.
Braun-Brumfield was printer and binder.
Cover design by Sharon Martin DePass.

Library of Congress Cataloging-in-Publication Data

Vul'fson, I. I. (Iosif Isaakovich)
(Vibroaktivnost' privodov mashin razvetvlennoĭ i kol'tsevoĭ struktury. English)
Vibroactivity of branched and ring structured mechanical drives /
I. Vulfson; English edition editor Eugene Rivin
p. cm. -- (Applications of vibration)
Translation of: Vibroaktivnost' privodov mashin razvetvlennoĭ i kol'tsevoĭ struktury.
Bibliography: p.
Includes index.

1. Power transmission. 2. Machinery—Vibration. I. Rivin, Eugene I. II. Title. III. Series.

ISBN 0-89116-813-3
ISSN 0897-8301
TJ1051.V8513 1988 88-16580
621.8'11—dc19 CIP

CONTENTS

PREFACE

The development and perfection of modern machinery raise many complicated technical problems for engineers. Some of them are related to the tendency toward intensification of process and transport operations, which is accompanied by increasing operating velocities of machines and dynamic loads in drives.

The problem of reducing vibroactivity of machinery is very important. Mechanisms for motion transformation and for programmable displacement of operating members are playing a double role in the vibratory system: on the one hand being the source of disturbances for the drive of the machine, its foundation, and load-carrying structure; and on the other hand being critical objects subjected to vibration. When designing drives for machines, the problem of reducing vibroactivity should be solved both with the aim of eliminating emergency regimes and also for maintaining normal conditions of operating and servicing of the machine. Thus, dynamic analysis of drives should become a necessary stage of machine design.

This book addresses issues related to vibroactivity of branched and ring drives which are representing large-scale dynamic systems. In a drive of a branched structure there are one or several distributing (main) shafts from which a great number of mechanisms, accomplishing kinematic and power transmission connections with the operating members of the machine, are branching off. In the ring structure drives, identical actuators operating in a parallel scheme are used in order to displace a massive operating member. By means of such drives, it is possible to reduce vibration of the operating member, to take up clearances, and to distribute loads among the mechanisms in a desirable fashion. The specifics of such drives, besides structural peculiarities, are due to application in various machines of cyclic mechanisms (cam mechanisms, linkages, etc.) which predetermine variability of parameters of the vibratory systems being analyzed. This excludes the possibility of a direct application of methods and results available in the literature for systems with constant parameters.

This book tries to bridge this gap in technical literature and to generalize recent results. While some information on analytical methods for systems with variable parameters is given in the book, a detailed and rigorous account of these methods was impossible because of its small volume.

General approaches to the considered class of problems and the analysis of a number of reference models may be found in the author's monograph [11].

The algorithms presented in this volume allow a user-friendly, computer-based approach to carrying out variation in the wide range of structural and dynamic characteristics of the drives, resulting in the choice of the optimum solution.

Some attention is given to effective engineering estimations, which form the qualitative understanding of dynamic processes and a general notion of the anticipated results of the computational analysis.

CHAPTER

ONE

DYNAMIC MODELS OF DRIVES AND THEIR MATHEMATICAL DESCRIPTION

§1. GENERAL INFORMATION ON DYNAMIC MODELS OF BRANCHED AND RING STRUCTURES

Basic stages of dynamic analysis. Even at the present level of the development of mechanics of machines and computer facilities, a full description of all aspects of dynamic behavior of a machine's elements and of the dynamic processes taking place within it is impossible (and unnecessary as well). Therefore, the first stage of dynamic analysis is concerned with reasonable simplification of the original object, i.e., its replacement by a certain schematic, or model, in which one should try to represent the most substantial factors of the considered problem. Since in this case we speak of problems of machine dynamics, the term "*dynamic model*" will be used.

The degree of idealization of a real system in its representation by a dynamic model is largely determined by specific conditions. Nevertheless, it is possible to pick up a number of standard models useful for many drives, both according to the goals of dynamic analysis and to their ability to represent the most important dynamic characteristics of the object investigated. It should be noted, however, that the selection procedure of a dynamic model requires a definite level of knowledge and understanding of the qualitative pattern of the phenomena in question. In some cases, the structural features of the drive under investigation are such that they allow for construction of one or several dynamic models. In the more complicated cases, to develop a succesful model one should make some preliminary calculations or even an experiment.

Each dynamic model is represented by a model, i.e., a set of equations providing a mathematical description of the dynamic model. For selection of a dynamic model it is necessary to understand the problem in question and to know specific features of the original physical object, while for the construction of a mathematical model it is necessary to know the basics of mechanics, vibration theory, the theory of mechanisms and machines, and also, frequently, experimental laws used for the description of certain forces (for example, process-related forces, etc.).

The third stage of dynamic analysis is the *solution of equations* obtained in the second stage when constructing the mathematical model. In this stage, both analytical methods giving a clear qualitative view and reliable engineering estimations, and numerical methods utilizing modern computer facilities are used. The numerical-analytical methods, based on a judicious combining of both methods, are very promising. When choosing the method of problem solution, one should not make abstractions from the assumptions laid in the process of idealization of the object being investigated in the stage of model construction.

The solution obtained may be used for the fourth stage of dynamic analysis, namely, for the *optimal dynamic synthesis* of the drive. As applied to the vibratory systems of machines, this problem is of interest for reducing the vibroactivity of mechanisms or for a more effective application of vibration in production processes. The development of methods of the optimal dynamic synthesis is one of the most important problems of modern machine dynamics.

Depending of the degree of importance of the results of dynamic analysis, the enumerated stages may be carried out at various levels of precision, both in relation to the chosen dynamic models and to the methods of their investigation.

For validation of the results of dynamic calculations the *experiment* is used. A specially important task is to check the correctness of the dynamic model. If the model has been chosen correctly, we may rely on the analysis and not subject all its points to an experimental checking (for instance, every new version of the mechanism, the influence of a variation of the drive parameters, etc.). Together with natural experiments with the object itself, a physical modelling sometimes is performed which involves the system in a different scale.

Usually the experiment is of a local nature since the wide variation of parameters and the system structures are labor-intensive and expensive. Here, the guiding idea constituting the theoretical core of the investigation is of great importance. The attempt to reduce all the dynamic investigations to the experiments results in the accumulation of vast material which can be generalized, in the best case, only on an empirical level. When solving dynamic problems, such an approach can do much harm in some cases. For example, if the natural frequencies of the system are not known beforehand, at least approximately, then a test in which the angular velocity of the input member of the drive is changed with some uniform step could easily end up passing over the resonance zones and, consequently, missing the most dangerous dynamic conditions.

Dynamic models of branched drives. Let us first consider some general concepts. Any dynamic model, regardless of its structural peculiarities, consists of individual elements which may be specified as lumped parameters or be of distributed character. Since the original real physical object, due to the elasticity of its components, has an infinite number of degrees of freedom, representation of this object by means of the dynamic model with discrete elements transforms it into a system with a finite number of degrees of freedom. At this transformation the following assumptions are usually made: 1. *the inertial properties of the system are represented by masses or moments of inertia which are concentrated in points or cross sections;* 2. *these points or cross sections are connected by elastic, dissipative, or kinematic links not having inertial properties.*

In practical application of these assumptions, the most massive elements as well as the most compliant (i.e., the least rigid) sections of the kinematic chain are identified in a drive or mechanism. The inertial, elastic, and dissipative properties of the system are taken into account by the reduced values of the corresponding parameters [4, 8, 11, 39, 50]. The reduction procedure is based on *the condition of invariability (in the first approximation) of kinetic and potential energy of the system, as well as the fraction of energy which is removed from the mechanical system due to action of the dissipative forces.*

A typical dynamic model of a branched structure with discrete elements is presented in Fig. 1*a*, where J_i are the moments of inertia (or masses), c_i are stiffness coefficients, ψ_i are the coefficients of dissipation, Π_i is the kinematic ana-

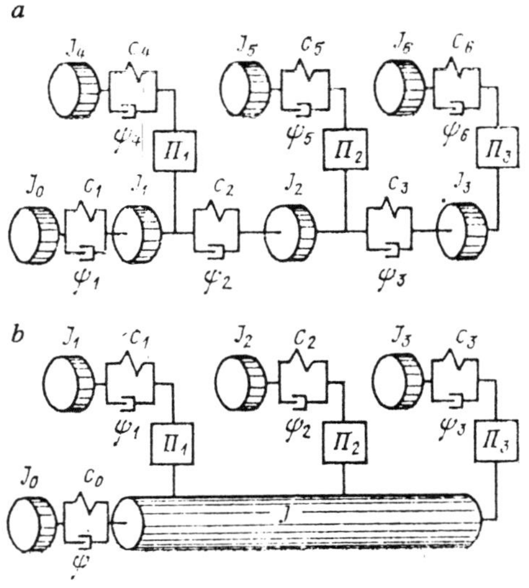

Figure 1. Typical dynamic models of branched drives.

log of the mechanism, which will be described below. It should be noted that supplementary intermediate branches can be added to any element, thus the model resembles the tree whose 'roots' are connected to the source of energy–the drive motor. Such a diagram is so typical to the overwhelming number of technological and transport machines and transfer lines that it does not require any additional illustrations. In some cases it is useful to represent separate links as systems with distributed parameters. This often results not only in a better accuracy of analysis, but to its simplification as well. This is done in cases when long main shafts are present (Fig. 1*b*), and also in schematization of operating members of technological machines such as long calenders, bars, beams, etc. The variants of models of the branched structure and their analysis are presented in Chapter 2.

Geometrical characteristics of an ideal mechanism. A distinctive characteristic property of dynamic models of drives and mechanisms is the geometric transformation of the absolute coordinate characterized by the position function Π. Usually, in any mechanism we can single out base members performing the translational or rotary (or oscillatory) motion. As a rule, such members are the driving and driven ones and possess the greatest moments of inertia or masses. The experience of engineering analyses shows that inertial and elastic properties of intermediate members performing more complex motions usually can be reduced to the base ones without any harm for the accuracy of calculations. With such an approach, the mechanism in the dynamic model may be represented as a set of the finite number of base members between which, in the general case, elements Π, c, ψ are located.

For example, in the slider-crank mechanism shown in Fig. 2*a* the connecting rod *2*, by means of the statical substitution of the mass, is split between the base members *1* and *3*. As a result, mechanism is represented in the form of the series connection of the ideal mechanism *1, 2, 3'* (Fig. 2*b*), and block c ψ, m, representing reduced (equivalent) elastic, dissipative, and inertial characteristics. The procedure is reflected in the model presented in Fig. 2*c*.

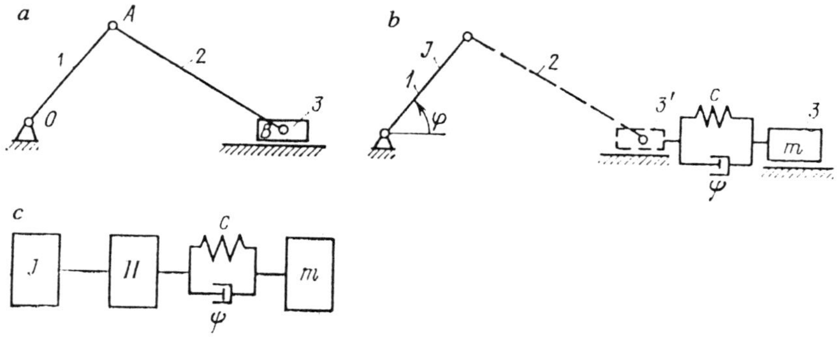

Figure 2. To the methods of construction of a dynamic model.

Suppose that φ_{j-1} and φ_j are the coordinates of the two successively located base members which are parts of the ideal mechanism, i.e., of an abstract mechanism in which the members do not deform and whose kinematic pairs do not have clearances. Then $\varphi_j = \Pi_j(\varphi_{j-1})$ is the position function of this mechanism and $\Pi_j^{(k)} = d^k\varphi_j/d\varphi_{j-1}^k$ is the kinematic transfer function of the order k. It is easy to prove that

$$\dot{\varphi}_j = \frac{d\varphi_j}{dt} = \Pi'_j(\varphi_{j-1})\dot{\varphi}_{j-1}$$
$$\ddot{\varphi}_j = \frac{d^2\varphi_j}{dt^2} = \Pi''_j(\varphi_{j-1})\dot{\varphi}^2_{j-1} + \Pi'_j(\varphi_{j-1})\ddot{\varphi}_j \tag{1}$$

The structure of expressions (1) shows that by using transfer functions, a clear distinction between geometrical and kinematic characteristics describing the motion of the member of the mechanism is achieved. In a specific case, for instance, in the gear train with constant transmission ratio, the position function Π is linear. As it follows from relationships (1), in this case $\dot{\varphi}_j = \Pi'_j\dot{\varphi}_{j-1}$; $\ddot{\varphi}_j = \Pi'_j\ddot{\varphi}_{j-1}$, while $\Pi'_j = u_{j,j-1}$ = const is the coefficient of proportionality between corresponding kinematic characteristics where $u_{j,j-1}$ is the transmission ratio. If in this case $\dot{\varphi}_{j-1}$ = const ($\ddot{\varphi}_{j-1} = 0$), then also $\dot{\varphi}_j$ = const ($\ddot{\varphi}_j = 0$).

With a nonlinear position function, typical for so-called *cyclic mechanisms* (cam mechanisms, linkages, etc.), the dynamic conditions of the operation appear to be more strenuous since even at $\dot{\varphi}_{j-1}$ = const we have $\ddot{\varphi}_j = \Pi''_j\dot{\varphi}^2_{j-1} \neq 0$.

Besides the relative geometrical characteristics introduced above, whose argument is the position of the preceding base member φ_{j-1}, the absolute geometrical characteristics connecting the member j with the member 1 are of interest. The connection between the absolute and relative geometrical characteristics is determined by the following relationships:

$$\varphi_k = \Pi_{1-k}(\varphi_1) = \Pi_{n-k}(\Pi_{1-n}(\varphi_1))$$
$$\Pi'_{1-k}(\varphi_1) = d\varphi_k/d\varphi_1 = \Pi'_{n-k}(\varphi_n)\Pi'_{1-n}(\varphi_1)$$
$$\Pi''_{1-k}(\varphi_1) = d^2\varphi_k/d\varphi_1^2 = \Pi''_{n-k}(\varphi_n)\Pi'^2_{1-n}(\varphi_1) + \Pi'_{n-k}(\varphi_n)\Pi''_{1-n}(\varphi_1)$$

Here, n is the arbitrary intermediate base member at $1 < n < k$; the prime means differentiation with respect to the argument appearing in the brackets.

The methods of determining geometric characteristics of specific mechanisms are presented in many monographs and textbooks [1, 11, 33].

Dynamic models of drives of the ring structure. By this term we mean the system of identical actuators forming kinematic chains with the common driving and driven members. So, for example, in the warp-knitting machines and stitch bonding machines the motion to the main operating members (the needle-holder, the plate, and the press) is supplied by several identical linkages or cam mechanisms whose driving members are installed on the main shaft (Fig. 3). The slay mechanisms of looms whose operating members are the com-

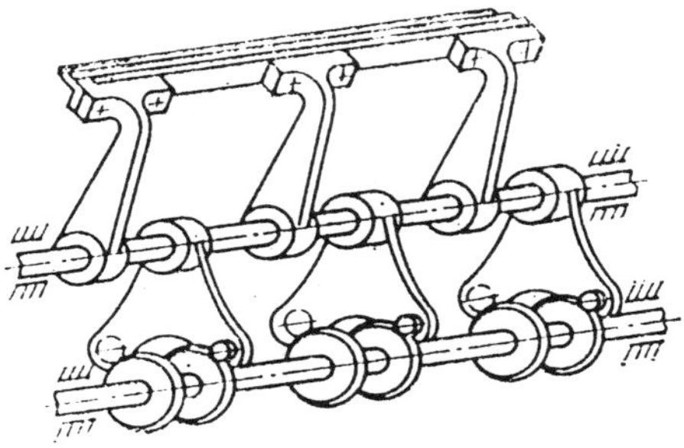

Figure 3. Driving mechanims of the operating member of a knitting machine.

mon driven members made in the form of long bars [27, 32], are of similar design. Such drives are widely used in the printing industry for moving heavy tables, carriages, and massive operating members (for example, book sawing machines, book cutters, etc.). In the mining and metallurgy industries a two-sided drive is used in the drive mechanisms of cold-rolling mills for pipes, in car dumpers, in the rotors of drying drums, etc.

In the examples presented, the realization of the technological process takes place in the zones of large dimensions, which leads to an increase of the overall dimensions of operating members; as a consequence, dynamic loads and vibroactivity in transmissions are also increasing. The duplication of mechanisms seeks to reduce loads in the kinematic chains as well as to reduce the vibration levels of the drive elements.

In some machines closed circuits are formed in the process of realization of technological operation when the original kinematic chain has an extra degree of mobility which is removed at the contact of the operating member with the workpiece being processed. In particular, such a situation arises in the clamping mechanisms of paper cutting machines, in rolling mills [40], etc.

The increased requirements for rigidity and chatter resistance imposed on copying (tracing) milling machines, NC machine tools, and other automatic machines are also realized by means of such drives [5, 25]. In this case the constant preload is developed in the kinematic chain, thus clearances which cause kinematic errors and disturbances of vibrational nature are eliminated.

A fragment of one of the typical ring structures is presented in Fig. 4*a* and its kinematic prototype in Fig. 4*b*. It should be noted that in many cases it is expedient to represent the driving and the driven members in the form of torsion subsystems with distributed parameters, as it is shown in Fig. 4*a* (for more details see Chapter 3). The characteristic property of all these drives of the ring structure is the static indeterminacy of the corresponding kinematic chains which shapes the approaches to the dynamic analysis of such systems. The studies carried out and the practical experience of operation of the ring structure drives show that the choice of the closed circuit does not by itself guarantee a successful solution of the dynamic problem. In some cases, installation of duplicating mechanisms can lead to an undesirable change of the spectrum of natural fre-

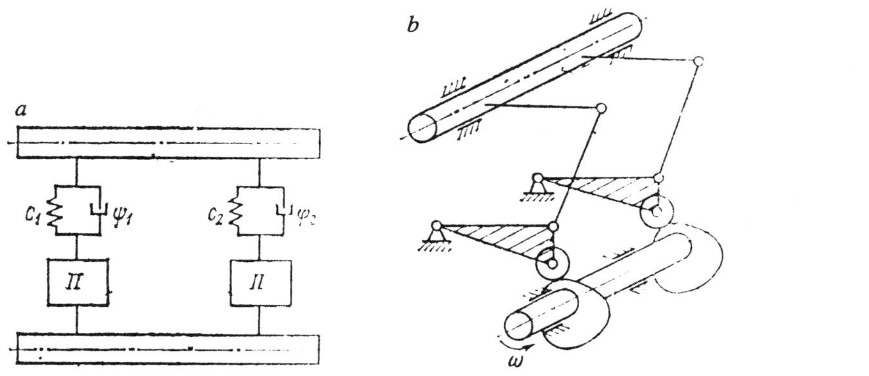

Figure 4. An example of the dynamic model of the mechanism with the ring structure.

quencies and, subsequently, to the increase of vibroactivity of the drive. The works [11, 12, 17, 21, 22] give general approaches to the solution of vibration problems in the ring structure drives, considering variability of the system parameters, which are further developed in this book (see Chapter 3).

Consideration of the dynamic characteristics of the motor. The dynamic models of branched and ring structure drives shown above should be, strictly speaking, complemented by an element corresponding to the *dynamic characteristic of the motor*. However, for analysts of steady-state regimes an easier way can be chosen when the angular velocity of the motor shaft is determined by relatively simple models. It does not affect, practically, the accuracy of the results. The model, consisting of the element *M*, which corresponds to the dynamic characteristic of an asynchronous electric motor or a d. c. motor, and the reduced moment of inertia *J*, is shown in Fig. 5*a*. Thus, only the inertial properties of the machine are represented in this model. It is possible, due to the fact that the motor is a low pass filter, that high frequencies of the mechanical system usually will have a negligible influence on the nonuniformity of angular speed of the rotor. Therefore, in the model presented in this stage we can either ignore the elastic-dissipative properties of the mechanical unit of the drive, or only take into consideration the most compliant elements, such as belt drives, elastic couplings, long transmissions, etc. (Fig. 5*b*). Through analysis of such a model we can find the coordinate $\varphi_0(t)$ defining in a first approximation the motion of the input member of the drive. For small nonuniformity of rotation, good results can also be obtained by the iteration method [31], when in the first stage the mechanical drive is investigated at $\varphi_0 = wt$, where w = const is the nominal angular velocity of the motor. Then the torque acting on the motor is

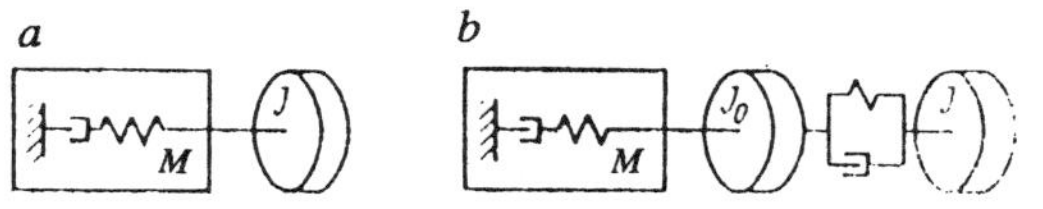

Figure 5. Dynamic models of a drive with a motor.

determined and, considering its dynamic characteristic, the vibratory components of angular velocity are determined. As a next step, corrections, including the nonuniform velocity of the rotor, are introduced into the calculations for the drive. Such an approach gives a clear picture of the influence of the nonuniformity of the machine operation of the dynamics of the drive. In view of the fact that influence of the dynamic characteristics of motors is amply covered in the technical literature [7, 8, 11, 31), in this book the function $\varphi_0(t)$ is assumed to be known.

§2. DERIVATION OF DIFFERENTIAL EQUATIONS FOR MECHANICAL DRIVES

Application of the second-kind Lagrange Equations. One possible method of compiling differential equations for the drive of the branched structure is demonstrated on a system consisting of the main shaft modelled as a three-disc vibratory system and of two cyclic mechanisms with nonlinear position functions of the driven members Π_1 and Π_2 (Fig. 6). The system has five degrees of freedom and should be described, accordingly, by five independent generalized coordinates. The generalized coordinates may be assigned in various ways. However, in this book we follow a definite order of their selection. Let us introduce the generalized coordinates as differences of the absolute coordinates of the respective inertial elements for the case when elasticity of links is considered, and on the assumption of absolutely rigid members. Then $q_1 = \varphi_1 - \varphi_0$; $q_2 = \varphi_2 - \varphi_0$; $q_3 = y_1 - \Pi_1(\varphi_0)$; $q_4 = y_2 - \Pi_2(\varphi_0)$ (see Fig. 6). If we assume that $\varphi_0 = \varphi_0(t)$, then the system would have nonstationary connections described by the functions Π_1 and Π_2. In such cases, however, for methodical reasons, at the stage of compilation of differential equations it is more convenient to use equations of a stationary connection, which is easy to do if in this stage the functional relation of the coordinate φ_0 with time is not revealed, and it is considered as one of the generalized coordinates $\varphi_0 = q_0$.

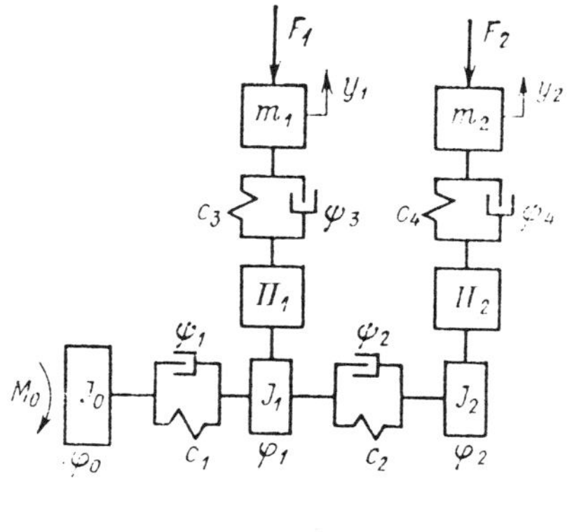

Figure 6. To the methods of mathematical description of dynamic models.

In a general case, the kinematic energy for the systems with stationary connections may be written as a quadratic form

$$T=\frac{1}{2}\sum_{i=0}^{H}\sum_{k=0}^{H}a_{ik}\dot{q}_i\dot{q}_k \tag{2}$$

where $a_{ik}=a_{ki}$ are the inertial coefficients depending on the generalized coordinates; $H+1$ is the number of degrees of freedom.

After substitution (2) into the second-kind Lagrange equation we obtain

$$\sum_{k=0}^{H}a_{jk}\ddot{q}_k+\frac{1}{2}\sum_{i=0}^{H}\sum_{k=0}^{H}\left(\frac{\partial a_{ji}}{\partial q_k}+\frac{\partial a_{jk}}{\partial q_i}-\frac{\partial a_{ik}}{\partial q_j}\right)\dot{q}_i\dot{q}_k$$
$$+\frac{\partial V}{\partial q_j}=Q_j \qquad (j=0,\ 1,\ \ldots,\ H) \tag{3}$$

where V is potential energy; Q_j is non-conservative generalized force; j is the ordinal number of the equation.

It should be emphasized that all the members containing $\dot{q}_i\dot{q}_k$ cannot be considered as small parameters of a second order, since among the generalized coordinates we have the coordinate q_0, which is not a small quantity.

The expression in brackets multiplied by the coefficient 0.5 is called the symbol of Christophel of the first kind [34] and has the abbreviated designation $[i, k, j]$. The coefficients $[i, k, j]$ in the general case have rather cumbersome expressions and require special attention during derivations. The special form of the second kind Lagrange equations with redundant coordinates has a definite advantage in this class of problems since these members are absent in the explicit form [11, 23, 34]. However, in this case this form is not used because Eqs. (3) at the chosen way of the assignment of generalized coordinates simplified as

$$\sum_{i=0}^{H}a_{0i}\ddot{q}_i+\partial V/\partial q_0=Q_0+P_0 \qquad (j=0) \tag{4}$$

$$a_{j0}\ddot{q}_0+a_{jj}\ddot{q}_j+\partial V/\partial q_j=Q_j+P_j \qquad (j=1,\ \ldots,\ H) \tag{5}$$

Here, $P_i=-\,[a_{ii}\Pi''_{0-i}(q_0)\dot{q}_0^2+0.5\dot{\varphi}_i^2\partial a_{ii}/\partial\varphi_i]$; $i=0, 1, \ldots, H$; $\varphi_i=\Pi_{0-i}(q_0)+q_i$ is the absolute coordinate corresponding to the element a_{ii}; the prime in the kinematic transmission functions means differentiation with respect to the argument in brackets. It is easy to make sure that the function P_i describes the inertia force reduced to the element i in the relative motion at $\dot{q}_0=\text{const}$.

Since $\Pi'_{0-0}=1$; $\Pi''_{0-0}=0$, we have $P_0=-0.5\dot{q}_0^2\partial a_{00}/\partial q_0$. In this case

$$T=0.5\{J_0\dot{q}_0^2+J_1(\dot{q}_0+\dot{q}_1)^2+J_2(\dot{q}_0+\dot{q}_2)^2$$
$$+m_1[\Pi'_1(q_0)\dot{q}_0+\dot{q}_3]^2+m_2[\Pi'_2(q_0)\dot{q}_0+\dot{q}_4]^2\} \tag{6}$$

Grouping the terms at $\dot{q}_i\dot{q}_k$ ($i, k = 1, 2, 3, 4$) and equating the corresponding coefficients in (2) and (6), we obtain $a_{00} = J_0 + J_1 + J_2 + m_1\Pi_1'^2 + m_2\Pi_2'^2$ $a_{11} = J_1$; $a_{22} = J_2$; $a_{33} = m_1$; $a_{44} = m_2$; $a_{01} = J_1$; $a_{02} = J_2$; $a_{03} = m_1\Pi_1'$; $a_{04} = m_2\Pi_2'$; the other coefficients are equal to zero.

Since $\varphi_1 = q_0 + q_1$; $\varphi_2 = q_0 + q_2$; $\varphi_3 = y_1 = \Pi_1(q_0) + q_3$; $\varphi_4 = y_2 = \Pi_2(q_0) + q_4$, have $\Pi_{0-1} = q_0$; $\Pi_{0-2} = q_0$; $\Pi_{0-3} = \Pi_1(q_0)$; $\Pi_{0-4} = \Pi_2(q_0)$. And $P_1 = 0$; $P_2 = 0$; $P_3 = -m_1\Pi_1''\dot{q}_0^2$; $P_4 = -m_2\Pi_2''\dot{q}_0^2$.

The potential energy of one linear elastic element is equal to $\Delta V_i = 0.5c_i\,\Delta\varphi_i^2$, where c_i is the coefficient of rigidity; $\Delta\varphi_i$ is elastic deformation. For the considered drive we have

$$V = 0.5\{c_1q_1^2 + c_2(q_2 - q_1)^2 + c_3[\Pi_1(q_0) + q_3 - \Pi_1(q_0 + q_1)]^2 + c_4[\Pi_2(q_0) + q_4 - \Pi_2(q_0 + q_2)]^2\} \tag{7}$$

Hence

$$\partial V/\partial q_0 = c_3[\Pi_1(q_0) + q_3 - \Pi_1(q_0 + q_1)] + c_4[\Pi_2(q_0) + q_4 - \Pi_2(q_0 + q_2)]$$

$$\partial V/\partial q_1 = c_1q_1 - c_3\Pi_1'(q_0 + q_1)[\Pi_1(q_0) + q_3 - \Pi_1(q_0 + q_1)] - c_2(q_2 - q_1)$$

$$\partial V/\partial q_2 = c_2(q_2 - q_1) - c_4\Pi_2'(q_0 + q_2)[\Pi_2(q_0) + q_4 - \Pi_2(q_0 + q_2)]$$

$$\partial V/\partial q_3 = c_3[\Pi_1(q_0) + q_3 - \Pi_1(q_0 + q_1)]$$

$$\partial V/\partial q_4 = c_4[\Pi_2(q_0) + q_4 - \Pi_2(q_0 + q_2)]$$

In order to find the nonconservative generalized forces Q_j, the equation of work of external forces on virtual displacements should be formed. If we limit ourselves to two technological forces F_1 and F_2 and the driving torque M_0, then

$$\delta W = M_0\delta\varphi_0 - F_1\delta y_1 - F_2\delta y_2 = M_0\delta q_0 - F_1\delta[\Pi_1(q_0) + q_3] - F_2\delta[\Pi_2(q_0) + q_4]$$
$$= [M_0 - F_1\Pi_1'(q_0) - F_2\Pi_2'(q_0)]\delta q_0 - F_1\delta q_3 - F_2\delta q_4$$

Hence, $Q_0 = M_0 - F_1\Pi_1'(q_0) - F_2\Pi_2'(q_0)$; $Q_1 = Q_2 = 0$; $Q_3 = -F_1$; $Q_4 = -F_2$. In a similar fashion, the other forces, such as dissipative forces, may be taken into consideration.

Now, all the components of the system of the differential equations (4) and (5) are obtained. If $\varphi_0 = q_0(t)$ is the known function of time (for instance, $\varphi_0 \approx \omega t$), then the differential equation (4) at $j = 0$ should be separated. Solving equations (5), first we find $q_1, ..., q_H$, and after the substitution of these functions into Eq. (4) we determine the unknown torque M_0. As it has been said, it is possible, in principle, to consider the motor within the unified model with the drive. In order to do it, the set of equations should be complemented by the equation of the dynamic characteristics of the motor connecting the driving torque M_0 with the angular velocity q_0 [7, 11, 31].

When the members of the drive are absolutely rigid $q_1 \equiv 0, \ldots, q_H \equiv 0$, then (4) coincides with the known relationship $J_0 \ddot{q}_0 + 0.5\dot{q}_0^2 \, \partial J_0/\partial q_0 = M_0 - M_c$, where M_c is the resistance torque reduced to the shaft of the motor [1, 33].

Linearization in the vicinity of the current value of the ideal phase angle φ_0. As it has been shown above, any coordinate in the absolute motion φ_i is a combination of the 'gross' coordinate $\Pi_{0-i}(q_0)$, realizing one degree of freedom of the absolutely rigid drive, and the 'small' coordinate q_i, whose ensemble corresponds to H degrees of freedom of the vibratory system. Here, the set of differential equations is nonlinear, since the generalized coordinates and their derivatives enter as arguments of nonlinear functions. However, by means of dynamically unimportant simplifications this system can be reduced to a set of linear differential equations with variable coefficients. This low price is paid for a possibility to use the *principle of superposition*. This principle allows to describe vibrations from the sum of forces by the sum of solutions from each force taken separately, or the solution of the system of heterogeneous equations form as the sum of solutions of the homogeneous system and the particular solution of the heterogeneous one. It is difficult to overemphasize the importance of this principle in the engineering analyses. In this case the price for it is negligible errors which are fully within the physical rigor of the construction of solutions and which correspond to the authenticity of all other initial information.

Represent now the geometrical characteristics of the drive Π_i, Π_i', Π_i'' in the form of the following truncated Taylor series [11, 23]: $\Pi_i^{(v)}(q_0 + q_i) \approx \Pi_{i*}^{(v)} + \Pi_{i*}^{(v+1)} q_i$, where v is the number of the derivative; the asterisk here and below corresponds to the argument $\varphi_0 = q_0$. It is obvious that this procedure requires continuity and differentiability of the function $\Pi_i^{(v)}$. If this condition is violated, the angle φ_0 must first be broken up into segments within which these requirements are satisfied, and then the solutions are matched on the boundaries of the segments [11].

The linearization carried out in the vicinity of the current value of the ideal phase angle should not be mixed with such a linearization when the nonlinear function on a selected segment is replaced by a linear one. In this case *all the nonlinear functions $\Pi_i^{(v)}$ retained their nonlinear properties relative to the gross coordinate q_0* and only small deformations q_i entered the corresponding expressions in a linear fashion. The data from several analyses show that even in rather stressed dynamic regimes the values of the further terms of Taylor series are such that they do not justify further refinement.

Using this method on the system (5), we obtain at $q_0 = \varphi_0(t)$

$$a_{jj}^*(t)\ddot{q}_j + \sum_{i=1}^{H} c_{ji}^*(t) q_i = Q_j - [a_{j0}^*(t)\ddot{q}_0 \tag{8}$$

$$+a_{jj}(t)\Pi''_{0-j*}\dot{q}_0^2+0.5(\partial a_{jj}/\partial q_0)_*\,(\dot{q}_0^2+2\dot{q}_j\dot{q}_0)] \qquad (j=1,\ \ldots,\ H)$$

where $c^*_{ik}=c^*_{ki}$ are quasielastic coefficients (with accuracy up to the performed linearization), appearing in the expression of the potential energy

$$V^*=0.5\sum_{i=1}^{H}\sum_{k=1}^{H}c^*_{ik}\,q_iq_k$$

Let us now illustrate the determination of quasielastic coefficients for the drive considered. After linearization, expression (7) becomes $V^*=0.5\,[c_1q_1^2+c_2(q_2-q_1)^2+c_3\,(q_3-\Pi'_{1*}q_1)^2+c_4\,(q_4-\Pi'_{2*}q_2)^2\,]=0.5\,[(c_1+c_2+c_3\Pi'^2_{1*})\,q_1^2+(c_2+c_4\,\Pi'^2_{2*})\,q_2^2+c_3q_3^2+c_4q_4^2-2c_2q_1q_2-2c_3\Pi'_{1*}q_1q_3-2c_4\Pi'_{2*}q_2q^4]$. From this $c_{11}=c_1+c_2+c_3\,\Pi'^2_{1*}$; $c_{22}=c_2+c_4\Pi^{12}_{2*}$; $c_{33}=c_3$; $c_{44}=c_4$; $c_{12}=-c_2$; $c_{13}=-c_3\Pi'_{1*}$; $c_{24}=c_4\Pi'_{2*}$, and the remaining coefficients are equal to zero. Since $c_{0k}=0$, the coordinate q_0 now does not enter into the expression for the potential energy V^*.

It should be noted that when $a_{ji}\neq$ const in the right part of the differential equation (8), the gyroscopic component proportional to q_j appears. Besides, the equivalent linear resistance forces (see §4, as well as [4, 11, 37]) can also be represented as proportional to generalized velocity in the first power.

In the following chapters, for the linearized (in this sense) systems, another approach would also be used—the apparatus of modified transfer matrices, whose application allows to exclude, in an explicit form, the procedure of the formation of sets of differential equations. Besides, this procedure will be used for analyses of the drives that include elements with distributed parameters.

§3. ANALYSIS OF DYNAMIC MODELS WITH VARIABLE PARAMETERS BY METHOD OF A CONDITIONAL OSCILLATOR

Single degree of freedom system. As it follows from (8), a following second order differential equation serves as a mathematical model for the simplest systems with variable parameters and a single degree of freedom

$$a^*(t)\ddot{q}+\dot{a}^*(t)\dot{q}+c^*(t)q=F(t)+R(q,\ \dot{q}) \tag{9}$$

Here, $R=-|R|\operatorname{sgn}\dot{q}$ is the dissipative force, which in the first approximation can be represented as the equivalent force of the linear resistance $R=-b\dot{q}$.

Let's transform Eq. (9) to the following form

$$\ddot{q}+2[n_0(t)+n_1(t)]\dot{q}+k^2(t)q=W(t) \tag{10}$$

where $n_0=b/(2a^*)$; $n_1=\dot{a}^*/(2a^*)$, $k^2=c^*/a^*$; $W(t)=F/a^*$

For solving differential equations with variable parameters, various methods can be used [35, 36, 45]. In this book the solution of such equations will be based on the *method of a conditional oscillator*. This method is commonly used to solve applied problems of mechanism dynamics [11, 18, 19]. According to this method

$$q = A_0 \exp\left[-\int_0^t n_0 dt\right]\sqrt{\frac{a^*(0)\,\Omega(0)}{a^*(t)\,\Omega(t)}}\cos\left[\int_0^t \Omega dt + \gamma\right]$$
$$+\frac{1}{\sqrt{\Omega(t)}}\int_0^t \frac{W(u)}{\sqrt{\Omega(u)}}\exp\left[-\int_u^t n(\xi)d\xi\right]\sin\left[\int_u^t \Omega(\xi)d\xi\right]du \tag{11}$$

Here, the amplitude of tree vibrations A_0 and the initial phase γ are defined by the initial conditions, and the function $\Omega(t)$ must satisfy the equation of the conditional oscillator

$$\ddot{z} - 0.5\,\dot{z}^2 + 2\Omega_*^2 e^{2z} = 2p^2(t) \tag{12}$$

where Ω_* is the arbitrary parameter having dimension of frequency and having the meaning of the normalizing multiplier; $z = \ln(\Omega/\Omega_*)$; $p^2(t) = k^2 - n^2 - n$ (in the problems of mechanism dynamics usually $|n^2 + n| << k^2$ which allows to take $p \approx k$).

Depending on the character of change of the function $p^2(t)$, it is possible to construct the approximate, and in some cases exact, solution of Eq. 12 [11, 18, 19].

Consider now some special cases important for applications.

1. Slow parameter change. In this case the parameter change of the model for one time-averaged period of free vibrations may be considered to be small in comparison with their average values in this period (one should not mix this case with small parameter changes relative to the average values over the long periods of time, such as a kinematic cycle of the mechanism). The inequality $|0.5\ddot{p}/p^3 - 0.75\,(\dot{p}/p^2)^2| \leq 0.2$ may serve as a quantitative criterion for this case.

Here, the condition $p \approx \Omega$ and the solution (11) agree with WKB-approximation of the first order [45].

2. Case. $|z| < 1$. This case is very common in practical applications since it corresponds to the sufficiently large range of change of 'natural' frequency, which may be estimated as $p_{max}/p_{min} \leq 4$. As it is shown in [11, 19], in this case the conditional oscillatior possesses pronounced linear properties. So, for instance, when $|z| < 1$, the main frequency of free vibrations of the conditional oscillator does not deviate more than 3.5 per cent from the mean value. This condition allows, in the present case, for carrying out the linearization of the co-efficients of Eq. (12), after which it takes the form

$$\ddot{z} + 4\bar{p}^2 z = 2(p^2 - \bar{p}^2) \tag{13}$$

where $\bar{p}^2 = \Omega_*^2$ is the mean value of the function $p^2(t)$.

Eq. 13 has an exact analytical solution.

Let us illustrate now the possibility of the parametrical excitation in the zone of the main parametric resonance. Suppose, for example, that pulsation of the function $p^2(t)$ takes place near the mean value $\bar{p}^2$. Pulsation frequency is ω; and $p^2 = \bar{p}^2(1 - \varepsilon \cos \omega t)$, where ε is the depth of the pulsation. Then from (13), $\ddot{z} + 4\bar{p}^2 z = -2\varepsilon \cos \omega t$. It is obvious that the conditional oscillator resonates at $\omega = 2\bar{p}$, which corresponds to the main parametric resonance of the original system. The amplitude build-up of the conventional oscillator is a necessary, though insufficient, condition for dynamic instability of the original system. On the other hand, it may be argued that the condition of a limited value of the variable amplitude in the first component of the solution (11) is sufficient (but not necessary) for dynamic stability. These conditons will be considered further.

3. Families of exact solutions. As it has been shown in [11, 18, 19], if the function $p^2(t)$ is piece-meal constant, then the equation of the conditional oscillator (12) has an exact analytical solution. Besides this case, exact solutions can be constructed for the whole families of functions $p^2(t)$ possessing certain properties.

Analytical method of construction of solution for calculating steady-state conditions. Let us introduce dimensionless time $\Phi = \bar{p} \int_0^t e^z \, dt'$ (here, $\bar{p}^2$ is the mean value of $p^2(t)$), and a new variable,

$$v = q \exp\left[\int_0^t n_1(t)\,dt + 0.5z \right] \tag{14}$$

In the new coordinates, equation (10) takes the form of a differential equation with constant coefficients

$$\frac{d^2 v}{d\Phi^2} + 2\delta \frac{dv}{d\Phi} + (1 + \delta^2)v = L(\Phi) \tag{15}$$

where $\delta = n_0 \bar{p}$; $\quad L = W \bar{p}^{-2} \exp\left[\int_0^t n_1(t)\,dt - 1.5z \right]$

In the non resonance conditions, both function $n_1(t)$ and forced vibrations of the conditional oscillator $z(t)$ are periodic functions of the period $\tau = 2\pi/\omega$, where ω is the mean value of $\dot{q}_0$. It is obvious that, in this case, function $L(\Phi)$ is also periodical, and the dimensionless value of the period of this function is equal to $\Phi(\tau) \approx \bar{p}\tau = 2\pi\bar{p}/\omega$. The dimensionless time Φ is a monotonously increasing function of time since $d\Phi/dt = \Omega(t) = \bar{p}e^z > 0$.

In practical cases, as it already has been mentioned, the condition $|z| < 1$ is usually satisfied. Therefore, forced vibrations of the conditional oscillator can be determined from the simple equation (13). In many cases there is an additional simplification due to a possibility of substituting several members of expansion into a Maclaurin series ($e^z \approx 1 + z + 0.5z^2$) instead of e^z in the subintegral expression for $\Phi(t)$.

Periodic function $L(\Phi)$ can be presented as a Fourier series for the argument Φ.

$$L(\Phi) = L_0 + \sum_{j=1}^{\infty} (L_{cj} \cos j\tilde{\omega}\Phi + L_{sj} \sin j\tilde{\omega}\Phi) \tag{16}$$

where

$$L_0 = \frac{1}{\Phi(\tau)} \int_0^{\Phi(\tau)} L(\Phi) d\Phi = \frac{\bar{p}}{\Phi(\tau)} \int_0^{\tau} L(t) e^{z(t)} dt$$

$$L_{cj} = \frac{2}{\Phi(\tau)} \int_0^{\Phi(\tau)} L(\Phi) \cos j\tilde{\omega}\Phi d\Phi$$

$$= \frac{2\bar{p}}{\Phi(\tau)} \int_0^{\tau} L(t) e^{z(t)} \cos j\tilde{\omega}\Phi(t) dt$$

$$L_{sj} = \frac{2}{\Phi(\tau)} \int_0^{\Phi(\tau)} L(\Phi) \sin j\tilde{\omega}\Phi d\Phi$$

$$= \frac{2\bar{p}}{\Phi(\tau)} \int_0^{\tau} L(t) e^{z(t)} \sin j\tilde{\omega}\Phi(t) dt$$

$$\tilde{\omega} = 2\pi/\Phi(\tau) \approx 2\pi/(\bar{p}\tau)$$

Having solved the differential equation (15) with excitation (16), and considering (14), we obtain

$$q = e^{-\int_0^t n_1(t)dt - 0.5z} \left[L_0 + \sum_{j=1}^{\infty} \frac{L_j \sin(j\tilde{\omega}\Phi(t) + \alpha_j - \Delta_j)}{\sqrt{(1 - j^2\tilde{\omega}^2)^2 + 4j^2\tilde{\omega}^2\delta^2}} \right] \tag{17}$$

where $\delta = n_0/\bar{p}(\delta^2 \ll 1)$; $L_j = \sqrt{L_{cj}^2 + L_{sj}^2}$; $\sin\alpha_j = L_{cj}/L_j$; $\cos\alpha_j = L_{sj}/L_j$

$\Delta_j = \mathrm{arctg}((2j\tilde{\omega}\delta/(1 - j^2\tilde{\omega}\alpha)$ $\qquad (j\tilde{\omega} \neq 2)$

This method is especially convenient at sufficiently smooth functions $p(t)$ and $W(t)$. It is usually the case in linkages when only a small number of terms can be retained in a Fourier series.

For sharp changes of the functions $W(t)$ and time histsories of motion having dwells, for which cam or step mechanisms are usually employed, the closed form solution is preferable [11]. Besides, an analytical method of formating such a solution, whose adventage is a clear identification of the sources of vibroactivity, as well as a combination numerico-analytical method presented below, are possible.

Numerico-analytical method of constructing a closed form solution for steady-state regimes. In analysis of steady-state regimes, application of purely numerical methods of calculation often causes large cumulative errors due to a great number of integration steps. The numerico-analytical method of constructing the solution does not have this shortcoming. The method consists of several stages.

1. Integration of the differential equation of the conditional oscillator (12) with zero initial conditions. As a normalized value Ω_*, it is convenient to take the average value for the cycle τ of the "natural" frequency, i.e., $p(t)$ or $\Omega_* = \bar{p}$. At this stage we calculate $z(t)$, $z(\tau)$, $\Phi(\tau) = \bar{p}\int_0^{\tau} e^z\,dt$, $\Omega(\tau) = pe^{z(\tau)}$.

2. The determination of the particular solution Y by the numerical integration of the initial differential equation (10) with zero initial conditions*. At this stage, $Y(\tau)$ and $\dot{Y}(\tau)$ are determined.

3. The determination of initial conditions corresponding to the steady-state regime. First we find $\xi = \sqrt{\xi_1^2 + \xi_2^2}$ where $\xi_1 = Y(\tau)$, $\xi_2 = \Omega^{-1}(\tau)\,[\dot{Y}(\tau) + 0.5\dot{z}(\tau)Y(\tau)]$. Then we determine the coefficient of disturbance accumulation μ taking into consideration the vibrations excited in the previous cycles [11]

$$\mu = \frac{1}{\sqrt{1 - 2e^{-\vartheta N}\cos 2\pi N + e^{-2\vartheta N}}} \tag{18}$$

where $N = \Phi(\tau)/\omega$; ϑ is the logarithmic decrement.

The plot of the coefficient μ is presented in Fig. 7.

The final expressions for the initial conditions in the arbitrary cycle of the steady-state motion regime have the form:

$$q_0 = \mu\xi\sin\gamma^0 \qquad \dot{q}_0 = \mu\xi\bar{p}e^{z(\tau)}\cos\gamma^0 \tag{19}$$

where $\gamma^0 = \gamma + \arcsin\,[\mu e^{-\vartheta N}\sin 2\pi N]$; $\sin\gamma = \xi_1/\xi$; $\cos\gamma = \xi_2/\xi$

4. The numerical integration of the original differential equation with the initial conditions (19) (see the footnote at stage 2).

In this method, only certain intermediate functions calculated in the limited interval of time are determined by numerical integration. Conditions of periodicity are inserted in the solution by an analytical method, specifically, by means of the method of the conditional oscillator. The latter fact has a substantial effect on the solution accuracy and allows for obtaining effective engineering estimates as well. In particular, Eq. (18) allows for determination of the greatest and the smallest effects of the vibrations excited in the previous cycles. It is easy to see

* For numerical integration in the case of broken continuity of the inertial coefficient a^* at the moment t_j, it should be assumed for matching of the solution that $q(t_j - 0) = q(t_j + 0)$; $a^*(t_j - 0)\,\dot{q}\,(t_j - 0) = a^*(t_j + 0)\,\dot{q}\,(t_j + 0)$.

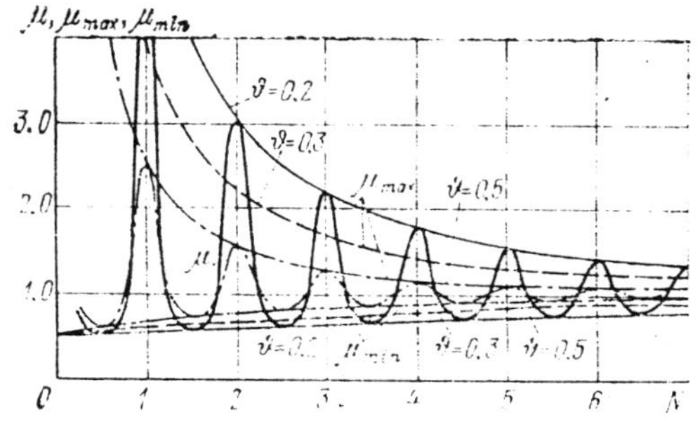

Figure 7. The plots of the coefficient of disturbance accumulation.

that when N is an integer, $\mu = \mu_{max} = (1 - e^{-\vartheta N})^{-1}$; if $2N$ is an odd number, then $\mu = \mu_{min} = (1 + e^{-\vartheta N})^{-1}$ (see Fig. 7).

If the parameters of the system are changing slowly, some simplifications of the method are possible. In this case there is no need to integrate the differential equation of the conditional oscillator since $\Omega = \bar{p}\, e^{z} \approx p$; $\Phi = \int_0^T p dt$; $z = \ln p/\bar{p}$; $N = \bar{p}/\omega$.

The multi-degree-of-freedom systems. As it has been pointed out above, the function n in the problems of dynamics of mechanisms has a negligible influence on the 'natural' frequency p and, at the same time, has a substantial effect on the amplitudes of vibrations. A similar picture is also revealed for systems with variable parameters which are represented by multi-degree-of-freedom models. Therefore, the frequency spectrum will be determined using the system of homogeneous differential equations

$$\mathbf{a}^*(t)\ddot{\mathbf{q}} + \mathbf{c}^*(t)\mathbf{q} = 0 \tag{20}$$

Here, $\mathbf{a}^*(t)$, $\mathbf{c}^*(t)$ are matrices of inertial and quasielastic coefficients after linearizing geometrical characteristics of mechanisms in the vicinity of the ideal value of the current phase angle (see §2).

The approximate solution according to the method of the conditional oscillator is

$$q_i = \sum_{r=1}^{H} a_{ir} \sqrt{\frac{\Omega_r(0)}{\Omega_r(t)}} \cos\left[\int_0^t \Omega_r(t)dt + \gamma_r\right] \tag{21}$$

In order to determine the function $\Omega_r(t)$ at the fixed values of t, it is necessary to solve first the system of equations

$$\det \| c_{ij}^* - a_{ij}^* p^2 \| = 0 \tag{22}$$

relative to the 'natural' frequency p*. The connection between p_r and Ω_r is described as before by the equation of the conditional oscillator

* When the number of degrees of freedom H is sufficiently great it is more convenient to use the apparatus of modified transfer matrices (see §5).

$$\ddot{z}_r - 0.5\dot{z}_r^2 + 2\Omega_r^2 e^{2z_r} = 2p_r^2(t) \tag{23}$$

with $\Omega_r = \Omega_{*r}\, e^{z_r}$ (Ω_{*r} is an arbitrary dimensional coefficient).

In the commonest case, when the parameters are changing slowly, we have $\Omega_r \approx p_r$.

The amplitudes of free vibrations a_{ir} at the fixed r constitute a mode of vibrations for whose description we introduce the non-stationary model coefficient $\alpha_{ir}(t) = a_{ir}/a_{rr}$. The modes of vibrations are assumed to be slowly changing, which corresponds to $\dot{\alpha}_{ir}/(\alpha_{ir} p_r) \ll 1$; $\ddot{\alpha}_{ir}/(\alpha_{ir} p_r^2) << 1$. It turns out, that here,

$$\ddot{q}_{ir} \approx -p_r^2(t) q_{ir} \tag{24}$$

for all combinations of i and r.

Relationship (24) is analogous to the corresponding result for the systems with constant coefficients but with the difference that $p_r \neq \text{const}$.

The nonstationary modal coefficients are determined from the following H sets of H-1 algebraic equations

$$\sum_{i=2}^{H} (c_{ji}^* - a_{ji}^* p_r^2)\alpha_{ir} = -(c_{j1}^* - a_{j1}^* p_r^2) \tag{25}$$

$$(j = 1, \ldots, H-1; \qquad r = 1, \ldots, H)$$

Here, j is the number of the equation in the set r, corresponding to the frequency p_r.

It should be noted that the elimination of the "fast" components of the function $\alpha_{ir}(t)$ when deriving Eq. (22) is not an obstacle for analyzing the influence of the fast components of the function $p_r^2(t)$ in the equation of the conditional oscillator. The latter is due to a small sensitivity of the "natural" frequencies to the changes of the modes of vibration [4, 39]. It should be noted that the Rayleigh method, as well as a number of other methods of the approximate determination of natural frequencies, is based on this property.

The convenience of this method in the problems of machine dynamics is largely due to the fact that the adopted form of solution, which corresponds to the method of the conditional oscillator, allows in accordance with (24) for use of formal procedures developed for systems with constant parameters. Thus, the procedure of analysis and calculation of vibratory systems of the mechanism becomes essentially a unified one.

§4. HIERARCHY OF MODELS. QUASINORMAL COORDINATES

As it already has been mentioned, the choice of an adequate dynamic model may be carried out on the basis of sufficiently clear intuitive notions or simple quantitative estimations of the elements of the vibratory system. However, in the considered class of cyclic mechanical systems of the branched and ring structure, á priori judgement on the degree of interrelation between individual vibratory circuits is not, as a rule, possible. In such a case, we may use the specific hierarchy of models, i.e., the successive considerations of dynamic models of two levels which we call "*global*" and "*local*" [13]. The global model allows to identify and, if necessary, to change the frequency spectrum of the system and the nonstationary modal coefficients. In this stage of the dynamic investigation it is possible to determine the degree of coupling between individual vibratory circuits of mechanisms and sensitivity of the system to the change of one or another parameter. In this stage, the solution is described by correlation (21), considering also (22) and (23); the dissipative factors and external disturbances are temporally not being considered.

On the basis of the local model, the solution is constructed taking into consideration both external disturbances and dissipative factors. The parameters of this model are determined on the basis of results of analysis of the global system using apparatus of quasinormal coordinates η_r. Here,

$$q_i = \sum_{r=1}^{H} \alpha_{ir}(t)\eta_r \qquad (i = 1, \ldots, H)$$

It can be shown that up to the values of the first-order of smallness, kinetic and potential energy are defined by relationships $T = 0.5 \sum_{r=1}^{H} J_r(t)\, \dot{\eta}_r^2$ and $V = 0.5 \sum_{r=1}^{H} c_r(t)\, \eta_r^2$, where the coefficients J_r and c_r are equal to, respectively,

$$J_r = \sum_{i=1}^{H} \sum_{j=1}^{H} a_{ij}^{*}\alpha_{ir}\alpha_{jr}; \qquad c_r = \sum_{i=1}^{H} \sum_{j=1}^{H} c_{ij}^{*}\alpha_{ir}\alpha_{jr} = p_r^2 J_r$$

The functions J_r and c_r may be qualified as the generalized mass and stiffness coefficient of the mode r.

At the adopted way of selecting the generalized coordinates of the vibratory system (see §2), parameters q_i are deviations from the program motion $a_{ij}^{*} = 0$ when $i \neq j$, $i \neq 0$. Then

$$J_r = \sum_{i=1}^{H} a_{ii}^{*}\alpha_{ir}^2 \tag{26}$$

In quasinormal coordinates, the system of differential equations becomes

$$J_r(t)\ddot{\eta}_r + (\bar{b}_r(t) + \tilde{b}_r(t))\dot{\eta}_r + c_r(t)\eta_r = P_r \qquad (r = 1, \ldots, H) \tag{27}$$

where $P_r = \sum_{i=1}^{H} Q_i^{*}\alpha_{ir}$ are new generalized nonconservative forces; Q_i^{*} are the right-hand side parts in the original system of the differential equations (8) (with exception of dissipative and gyroscopic components proportional to $\dot{q}_i$); $\bar{b}_r(t)$ and $\tilde{b}_r(t)$ are the coefficients taking into consideration the dissipative forces and gyroscopic components.

In the first approximation, dissipative forces can be considered with *an as-*

sumption that there is no dissipative coupling between various modes of vibration. It corresponds to the assumption that the energy transfer between the modes, caused by the nonlinear dissipative forces, is small. Then each mode of vibrations is associated with a relative dissipation $\psi_r = \Delta V_r / V_r$, where ΔV_r and V_r are the energy dissipated per cycle and maximum potential energy, respectively, corresponding to the mode r. If an elastic-dissipative element with the stiffness coefficient c_s and a relative dissipation ψ_s is confined between the cross sections whose modal coefficients are equal to α_{sr}^- and α_{sr}^+, respectively, then

$$\psi_r = c_r^{-1} \sum_{s=1}^{N} \psi_s c_s (\alpha_{sr}^+ - \alpha_{sr}^-)^2 \tag{28}$$

where N is the number of the elastic-dissipative elements. In this case the coefficient of the equivalent linear dissipation is equal to $\bar{b}_r = \psi_r c_r / (2\pi p_r) = \psi_r p_r J_r / (2\pi)$.

Enhancement of accuracy of solutions due to consideration of energy transfer between the different modes of vibrations and the nonlinear character of dissipative forces are considered in (9, 11, 37). In the linear approximation the following equivalent linear dissipation is due to the gyroscopic component:

$$\tilde{b}_r \approx \omega_0 \partial J_r / \partial \varphi_0 = \omega_0 \sum_{i=1}^{H} [\alpha_{ir}^2 \partial a_{ii}^* / \partial \varphi_0 + 2 a_{ii}^* \alpha_{ir} \partial \alpha_{ir} / \partial \varphi_0] \tag{29}$$

where $\omega_0 = d\varphi_0 / dt$.

At $a_{ii}^* = \text{const}$ in relationship (29), only the second term, which is associated with the variability of the modal coefficients, is retained.

For the calculations of the parameters of the local model, use of the quadratic forms of kinetic and potential energy is not obligatory. Usually, "natural" frequency p_r and nonstationary modal coefficients α_{ri} are determined by using modified transition matrices.

§5. MODIFIED TRANSITION MATRICES

Transition matrices for elements of a dynamic model. The apparatus of transition matrices is widely used for analysis of linear vibratory systems with constant parameters [4]. This apparatus, which is well adapted to calculation on computers, was developed in [11, 12, 13] for the problems of dynamics of mechanisms described by the system of differential equations with varying coefficients.* The possibility of expanding the application of transition matrices for models with varying parameters results from the fact that at the additional assumptions, which are realized in the method of the conditional oscillator, equality (24) is satisfied for each mode at free vibrations within the adopted assumptions. Ob-

* In order to distinguish the transition matrices when the parameters of the system are variable, they were named "modified." Henceforth, for the sake of brevity, this term will be applied only in cases when it is necessary to emphasize specifics of these matrices.

viously, we have the analogous relationships at p = const in systems with constant parameters.

Taking into consideration the specifics of the considered problems, we will limit ourselves to analysis of the single-connected systems. *The system connectivity shows the number of possible displacements of any cross section or, put another way, the number of reaction forces replacing action of one part on another when the system is dissected.*

For any mode of vibrations, the varying amplitude of free vibrations a_{i-1} and the amplitude of load (force or torque) Q_{i-1} undergo certain changes when passing through an element of the dynamic model j (inertial, elastic, or kinematic). It corresponds to the following matrix notation:

$$\begin{bmatrix} a_j \\ Q_j \end{bmatrix} = \begin{bmatrix} A_j & B_j \\ C_j & D_j \end{bmatrix} \begin{bmatrix} a_{j-1} \\ Q_{j-1} \end{bmatrix} \tag{30}$$

where a_{j-1}, a_j, Q_{j-1}, Q_j are amplitudes and amplitude loads in the cross sections j–1 and j, respectively.

The modified transition matrix Γ_j, whose elements are functions A_j, B_j, C_j, D_j, has a following property: det $\Gamma_j = 1$.

When using transition matrices, the definite rule of signs should be adhered to, namely: the reaction torque or the force on the "output" (the cross section j for the element j) are considered to be positive if their direction coincides with the positive coordinate direction; on "input" (the cross section j–1 for the element j) the rule of signs is opposite. The external loads are considered to be positive if they coincide with the adopted positive direction. With such a rule, it is not necessary to point out to which of the cross sections the reaction force or torque is applied. It reduces substantially the possibility of an error in the signs.

In the first three rows of Table 1, the components of the transition matrices are given for simple discrete elements of the dynamic model, namely, the elastic element with the stiffness coefficient c, the inertial one with the moment of inertia (or mass) J, and the kinematic analog Π. In the last case, after linearization of the function Π_j in the vicinity of the given coordinate of the programmed motion, the first transter function $\Pi'_j = d\Pi_j/d\varphi_{j-1}$ (where φ_{j-1} is the input coordinate) would be included in the transition matrix. At the constant transmission ratio $u_{j,j-1}$, $\Pi_j = u_{j,j-1}$. Also in the table are listed some cases of the series connection of simple discrete elements; the method of obtaining these matrices is given below. The last row of Table 1 corresponds to the transition matrix for the element with distributed parameters at torsional vibrations of the shaft segment. The following designations are used $\sigma = (GI\rho)^{-0.5}$; $\theta = p(t)\, l/g$; $g = \sqrt{GI/\rho}$, where G is the shear modulus, I is the polar moment of inertia of the shaft cross section; l is the length of the segment; ρ is the mass moment of inertia of the unit length of the shaft. When longitudinal vibrations of the bars are considered in the presented relationships, G should be replaced with the modulus of elasticity E, the moment of inertia I with the cross sectional area S, and ρ is the mass per unit length of the bar.

Table 1. Elements of transition matrixes for typical cases

Connection	A	B	C	D
c	1	c^{-1}	0	1
J	1	0	$-Jp^2$	1
Π	Π'	0	0	$1/\Pi'$
$c—J—\Pi$	Π'	$\Pi'c^{-1}$	$-Jp^2/\Pi'$	$(1-Jp^2c^{-1})/\Pi'$
$c—\Pi—J$	Π'	$\Pi'c^{-1}$	$-Jp^2\Pi'$	$-Jp^2\Pi'c^{-1}+1/\Pi'$
$J—c—\Pi$	$\Pi'(1-Jp^2c^{-1})$	$\Pi'c^{-1}$	$-Jp^2/\Pi'$	$1/\Pi'$
$J—\Pi—c$	$\Pi'-Jp^2/(c\Pi')$	$(c\Pi')^{-1}$	$-Jp^2/\Pi'$	$1/\Pi'$
$\Pi—J—c$	$\Pi'(1-Jp^2c^{-1})$	$(c\Pi')^{-1}$	$-Jp^2/\Pi'$	$1/\Pi'$
$\Pi—c—J$	Π'	$(c\Pi')^{-1}$	$-Jp^2/\Pi'$	$(1-Jp^2c^{-1})/\Pi'$
Element with distributed parameters	$\cos\theta$	$\sigma p^{-1}\sin\theta$	$-\sigma^{-1}p\sin\theta$	$\cos\theta$

Series connection of model elements. Suppose the dynamic model is represented as a series connection of n elements ($j = 1, ..., n$). By means of the series substitution of the matrices (30) arrive to

$$\begin{bmatrix} a_n \\ Q_n \end{bmatrix} = \begin{bmatrix} A_n & B_n \\ C_n & D_n \end{bmatrix} \cdots \begin{bmatrix} A_1 & B_1 \\ C_1 & D_1 \end{bmatrix} \begin{bmatrix} a_0 \\ Q_0 \end{bmatrix}$$

Thus, the transition matrix of a vibratory chain consisting of the series connection of elements is the product of matrices of these elements

$$\Gamma = \prod_{j=n}^{1} \Gamma_j \tag{31}$$

It should be noted that the sequence of cofactors in the product is opposite to the sequence of the corresponding simple elements since the matrix product does not have the property of commutativity.

Parallel connection of model elements. Suppose that n elements form a parallel connection. For each of the elements the following matrix equation is valid

$$\begin{bmatrix} a_{j\nu} \\ Q_{j\nu} \end{bmatrix} = \begin{bmatrix} A_{j\nu} & B_{j\nu} \\ C_{j\nu} & D_{j\nu} \end{bmatrix} \begin{bmatrix} a_{j-1,\nu} \\ Q_{j-1,\nu} \end{bmatrix} \qquad (\nu = 1, ..., n)$$

Since at the parallel connection the coordinates on the input and output for all elements are equal and, consequently, do not depend on v, we have $a_{jv} = a_j$ and $\omega_{j-1,v} = a_{j-1}$. Here,

$$\left.\begin{array}{l} a_j = A_{jv} a_{j-1} + B_{jv} Q_{j-1,v} \\ Q_{jv} = C_{jv} a_{j-1} + D_{jv} Q_{j-1,v} \end{array}\right\}$$

The resultant amplitude loads Q_{j-1} and Q_j now may be represented as

$$Q_{j-1} = \sum_{v=1}^{n} Q_{j-1,v} = -\varkappa_1 a_{j-1} + \varkappa_2 a_j$$

$$Q_j = \sum_{v=1}^{n} Q_{jv} = -\varkappa_2 a_{j-1} + \varkappa_3 a_j$$

where

$$\varkappa_1 = \sum_{v=1}^{n} A_{jv}/B_{jv}; \qquad \varkappa_2 = \sum_{v=1}^{n} B_{jv}^{-1}$$

$$\varkappa_3 = \sum_{v=1}^{n} D_{jv}/B_{jv}$$

The transition matrix Γ_j for the ensemble of parallel connected elements should comply with Eq. (30) at $A_j = \chi_1/\chi_2$; $B_j = \chi_2^{-1}$; $C_j = (\chi_1\chi_3 - \chi_2^2)/\chi_2$; $D_j = \chi_3/\chi_2$.

The methods of determination of "natural" frequencies and non-stationary modes by means of transition matrices. As it already has been mentioned in §4, the determination of "natural" frequencies and modes is performed on the basis of the global model. In the expanded form the connection between a_0, Q_0 and a_n, Q_n can be presented as

$$\left.\begin{array}{l} a_n = A(p)a_0 + B(p)Q_0 \\ Q_n = C(p)a_0 + D(p)Q_0 \end{array}\right\} \tag{32}$$

where A, B, C, and D are the elements of the transition matrix of the whole chain.

Depending on boundary conditions, consider now some particular cases.

1. "Input" is fixed, "output" is free, which corresponds to $a_0 = 0$ and $Q_0 = 0$. Since $Q_0 = a_n/B$, then $Q_n = Da_n/B = 0$. Hence, we obtain a formal frequency equation $D(p) = 0$.
2. "Input" is fixed, "output" is fixed ($a_0 = a_n = 0$); here $B(p) = 0$.
3. "Input" is free, "output" is free ($Q_0 = Q_n = 0$). In this case $C(p) = 0$.
4. "Input" is free, "output" is fixed ($Q_0 = 0$; $a_n = 0$). These conditions can be realized only when $A(p) = 0$.

After solution of the formal frequency equation by means of the transition matrices, it is easy to determine nonstationary modal coefficients. For this it is enough to take the amplitude in one of the cross sections as unity and, using the boundary conditions, to determine amplitudes in the other cross sections corresponding to the previously found root P_r. The values obtained are modal coefficients. The minus sign in the modal coefficient shows that the vibrations in the considered cross section and in the cross section where the modal coefficient was assumed to be unity are in opposite phases.

Let's determine, for example, $\alpha_{jr} = a_{jr}$ at $a_0 = 1$ (the cross section with the ordinal number j; $p = p_r$) for case 4:

$$\begin{bmatrix} a_{jr} \\ Q_{jr} \end{bmatrix} = \begin{bmatrix} A_j^*(p_r) & B_j^*(p_r) \\ C_j^*(p_r) & D_j^*(p_r) \end{bmatrix} \begin{bmatrix} 1 \\ 0 \end{bmatrix}$$

where A_j^*, B_j^*, C_j^*, D_j^* are the elements of the matrix

$$\Gamma_j^* = \prod_{s=j}^{1} \Gamma_s$$

It is obvious that $a_{jr} = \alpha_{jr} = A_j^*(p_r)$. This issue, as applicable to branched and ring systems, will be elaborated in Chapters 2 and 3.

Since p_r changes in time, the modal coefficients will also be changing. However, in the modal coefficients only slow components corresponding to the frequencly of the kinematic cycle ω should be retained.

In conclusion, it should be emphasized that despite the formal resemblance to the analogous procedure in the systems with constant parameters, in this case only a specific form of the particular solution of the set of homogeneous equations corresponding to the method of the conditional oscillator is considered.

§6. GENERAL INFORMATION ON METHODS OF REDUCING VIBROACTIVITY OF DRIVE QUASISTEADINESS CONDITIONS

Preliminary remarks. Nowadays, the problem of reducing vibroactivity is one of the most important problems of dynamic synthesis. The solution of this problem is developed in some directions, among which we should single out the following ones: the rational synthesis of geometrical characteristics of a mechanism possessing optimal properties proceeding from the chosen dynamic criterion [1, 11, 23, 33, 46]; the reduction of vibroactivity by a directed change of parameters of the vibratory system [11]; and the installation of special antivibration devices [9, 40, 42, 48].

The increased level of drive vibroactivity exhibits itself in the form of dynamic errors, i.e., distortions of the programmed motions of the drive elements. The big dynamic errors are developing due to the overloads which very often exceed substantially the level of the kinetostatic loads corresponding to the case of absolutely rigid members.

One should remember that the solution of a particular problem of vibroactivity reduction is often of a local character and sometimes is connected with the increased vibration level in the other part of the system. For example, balancing of the mechanism by counterweights can cause an increase of the variable component of the equivalent moment of inertia, which increases the drive vibroactivity. All this testifies that a systems approach to the problem is necessary.

In this section we are going to limit ourselves only to some directions for reduction of vibroactivity which are specific to the systems with variable parameters. These problems will be concretized further for the branched and ring structure drives (see Chapters 2 and 3). The general approach to this problem is presented in more detail in the monograph [11] and partially in [9].

Dynamic stability conditions on any time interval of the kinematic cycle (Quasisteadiness conditions). In §4 it has been shown that after the transformation to the quasinormal coordinates, the original system can be described by H differential equations of the form

$$\ddot{\eta}_r + 2n_r(t)\dot{\eta}_r + p_r^2(t)\eta_r = W_r(t) \tag{33}$$

The solution of this equation is described by expression (11) at $n_r = n_0 + n_1$; $\Omega_r = \Omega$. Suppose that the coefficients of the differential equation (33) are slowly changing functions. Then according to (11) and (27) the amplitude of free and accompanying vibrations excited due to various pulse disturbances is changing in time propotional to the function

$$Z_{rj} = [J_r(t)p_r(t)]^{0,5} \exp\left[-\int_{t_j}^{t} n_{r0}\,dt\right] \qquad (t \geqslant t_j) \tag{34}$$

It is easy to make sure that at constant parameters the function Z_{rj} is always decreasing and, therefore, that the free and accompanying vibrations are decaying. With changing parameters it may happen that $dZ_{rj}/dt > 0$, therefore the traditional decaying character of variation of these vibrations can be disturbed. This effect reveals a violation of the conditions of dynamic stability. In the similar case, the zone of the build-up is replaced by the zone of decay, therefore we do not experience the unlimited increase of the amplitudes typical, for example, to a parametric resonance. Nevertheless, under some unfavorable conditions the increase of amplitudes may become rather intensive. This calls for the dynamic synthesis to remove the possibility of the occurrence of such zones. The latter is easy to achieve by requiring $dZ_{rj}/dt < 0$. After the substitution of (34) into this condition we obtain, considering $n_{r0} \approx \psi_r p_r/(4\pi)$ where ψ_r is the coefficient of dissipation,

$$\psi_r > -2\pi p_r^{-1}(\dot{J}_r J_r^{-1} + \dot{p}_r p_r^{-1}) \tag{35}$$

In a similar way we can write conditions of dynamic stability for vibration velocity and vibration acceleration: $\psi_r > 2\pi p_r^{-1}(\dot{p}_r p_r^{-1} - \dot{J}_r J_r^{-1})$; $\psi_r > 2\pi p_r^{-1}(3\dot{p}_r p_r^{-1}$

$-\dot{J}_r J_r^{-1}$). From these three inequalities, usually, the last one happens to be determining.

It is possible to show that condition (35) can also be obtained by the direct Lyapunov method and is, consequently, the sufficient condition for absolute stability [11, 36]. Compliance with this condition removes also the possibility of the build-up in the zones of the main parametric resonances. Methods of damping of the combination resonances are presented in general in the monographs [6, 47] and, as applied to the problems of mechanisms, in [16].

Since at $Z_{rj}/dt < 0$ the system loses its most dangerous nonstationary properties, the obtained conditions may be treated as *conditions of quasisteadiness.*

In some cases, the quasisteadiness conditions as applied to individual modes of vibrations are satisfied by means of proper structural design of the drive [11] or by introducing special correcting linkages [14].

Sources of drive vibroactivity. With the assumed generalized coordinates the absolute coordinate φ_i of the arbitrary element of the drive i is described as

$$\varphi_i = \Pi_{i*} + q_i = \Pi_{i*} + \sum_{r=1}^{H} \alpha_{ir}\eta_r \tag{36}$$

where α_{ir} are non-stationary modal coefficients.

For the steady-state vibratory conditions we can show that

$$|\eta_r| \leq \mu_r \sum_{j=0}^{s-1} D_{rj}\tilde{Z}_{rj} + |\eta^*_{rs}| \tag{37}$$

where μ_r is the coefficient of accumulation of disturbances (see Eq (18)); $\tilde{Z}_{rj}(t) = Z_{rj}(t)/Z_{rj}(t_j)$; D_{rj} is the "jump" (see below); η^*_{rs} is the particular solution of Eq. (33) on the interval $t_{s-1} < t < t_s$.

Using Eq. (37) let's follow-up the sources of vibration excitation. The first group of terms is proportional to the amplitude of accompanying vibrations; the period of the kinematic cycle $\tau = 2\pi/\omega$ is broken up into s intervals within which the functions Π_{i*} and W_r and some of their derivatives do not have breaks of continuity. In the moment of time $t = t_j$ (the boundary of the intervals j–1 and j) accompanying vibrations are excited whose amplitude is proportional to the jump D_{rj} [9, 11] and

$$D_{rj} \approx \sqrt{\left(\Delta\eta_{rj} - \frac{\Delta W_{rj}}{p^2_{rj}} + \frac{\Delta \ddot{W}_{rj}}{p^4_{rj}}\right)^2 + p_{rj}^{-2}\left(\Delta\dot{\eta}_{rj} - \frac{\Delta \dot{W}_{rj}}{p^2_{rj}} + \frac{\Delta \dddot{W}_{rj}}{p^4_{rj}}\right)^2} \tag{38}$$

where $p_{rj} = p_r(t_j)$; $\Delta W^{(\nu)}_{rj}$ is the stepwise change of the corresponding derivative ν of the function W_r in the moment of time $t = t_j$; $\Delta\eta_{rj}$, $\Delta\eta^*_{rj}$ are the "initial" conditions arising at the breaks of continuity of the functions Π_* and Π'_* in the moment of time $t = t_j$.

It follows from Eq. (38) that it is necassary first of all to eliminate the jumps of $\Delta W_r^{(\nu)}$, especially when $\nu = 0$, and to make $\Delta\eta_{rj}$ and $\Delta\dot{\eta}_{rj}$ vanish. The appearance of these "initial" conditions is connected with the fact that when the external shocks are absent, the functions φ_i and $\dot{\varphi}_i$ must be continuous. Therefore, at the jumps $\Delta\Pi_{ij*}$ and $\Delta\Pi'_{ij*}$ in the moment $t = t_j$, according to (36), the stepwise changes $\Delta q_{ij} = -\Delta\Pi_{ij*}$ and $\Delta\dot{q}_{ij} = -\Delta\Pi'_{ij*}\,\omega$ occur, which would correspond to $\Delta\eta_{rj}$ and $\Delta\dot{\eta}_{rj}$ at the transformation to the quasinormal coordinates. The breaks of the derivatives of the function Π_{i*} of a higher order are reflected in the jumps $\Delta W_{rj}^{(\nu)}$. Consequently the function Π_{i*} must be continuous and differentiable at least three times.

It is possible to show that the sufficiently sharp changes of the function W_r also cause the jump effect. In order to avoid this effect for several lower frequencies ($r = 1, 2$) it should be assumed that $\Delta t_j > (2.5\text{–}4)\ T_{rj}$ where $T_{rj} = 2\pi / p_{rj}$ is the averaged value of the period of free vibrations on this interval; Δt_j is the time of increasing or decreasing of the function W_r from zero to extremum.

When condition (35) is satisfied we have a decaying vibratory process, therefore $\tilde{Z}_{rj} \ll 1$. The accumulation coefficient of disturbances takes account of accompanying vibrations excited in the previous cycles (see Eq. (18)). If we assume the permissible value of the coefficient $[\mu_r]$, then the condition

$$N_r = \bar{p}_r\,\omega \geq \frac{2}{\psi_r} \ln \frac{[\mu_r]}{[\mu_r] - 1}$$

must be satisfied; ψ_r is the coefficient of dissipation reduced to the mode r.

When $N_r\psi_r \geq 6$, then we have $0.96 < \mu_r < 1.04$.

The second component of (37), which corresponds to the particular solution, depends on the chosen motion time histories and process loads. A reduction of this component can be achieved by the rational synthesis of the function Π_{i*} as well as by means of special unloading devices [42].

Clearances. The dynamic effect due to the presence of clearances in the kinematic pairs is pronounced in various ways. Strictly speaking, the clearances disrupt the linearity of the vibratory system since at the displacements in the clearances the restoring force is equal to zero. Thus the clearance acts as a nonlinear element acting on the frequency properties of the system. However, in many cases the transition through the clearance occurs only several times during the period of the kinematic cycle in the zone of the sign change of the exciting force. In this case, at $N_r \gg 1$, ($p_r \gg \omega$), with the exception of the mentioned small "switching" zones, the system retains its linear properties and reacts on the clearance as to some pulse disturbances. As it has been shown in [11], in such cases the clearance is equivalent to the jump of the first transmission function $\Delta\Pi'_{j*}$, and $\Delta\Pi'_{j*} \approx \sqrt{2\Delta s\,|\Pi''_*(t_j)|}$ at $\Pi''_*(t_j) \neq 0$, and $\Delta\Pi'_{j*} \approx \sqrt[3]{4.5\Delta s^2 |\Pi'''_*(t_j)|}$ at $\Pi''_*(t_j) = 0$ and $\Pi'''_*(t_j) \neq 0$. Naturally, from the engineering point of view, when solving the problems of dynamic synthesis the criterion according to which it is possible to evaluate dynamic effects caused by the clearance would be of great interest.

For cyclic mechanisms one such criterion can be obtained directly from the relationships presented above as $\Delta\varepsilon_j/\varepsilon_{max} = \chi_j$, where $\Delta\varepsilon_j$ is the acceleration jump which is equivalent in its dynamic effect to the jump $\Delta\Pi'_{j*}$; ε_{max} is the maximum acceleration. It is easy to show that $\chi_j = N\Delta\Pi'_{j*} : |\Pi''|_{max}$, where $N = \bar{p}_1/\omega$, ω is the angular velocity, and $\bar{p}_1$ is the averaged lowest "natural" frequency. During the synthesis it should be required that $\chi_j \leq 0.1$; at $\chi_j \to 1$ the clearance causes the same additional accelerations of the system as a soft impact.

The next criterion can be obtained by the method of harmonic linearization. Let's consider the series connection of the element with a clearance equal to $2\Delta s$ and an elastic element with the stiffness coefficient c_0 (Fig. 8). Suppose that the force $F(t)$ is transferred over the chain. As it was shown in (2), for such nonlinearities the output process is only weakly sensitive to the format of the input action. Therefore, for linearization the force $F(t)$ may be described in a sufficiently crude approximation as $F(t) = F_0 + F_1 \cos \omega t$. After the harmonic linearization by force, we note that the combination of the clearance with the elastic element is equivalent to the conditional elastic element whose compliance is equal to $e_* = c^{-1} = e_0 + e_\Delta$ where $e_0 = c^{-1}$ and

$$e_\Delta = \begin{cases} \frac{4\Delta s}{\pi F_1}\sqrt{1-\left(\frac{F_0}{F_1}\right)^2} & \text{at} \quad F_1 > F_0 \\ 0 & \text{at} \quad F_1 < F_0 \end{cases}$$

Compare now the maximum energies E_* at $\Delta s \neq 0$ and E_0 at $\Delta s = 0$. Since $E_* = 0.5\,(F_0 + F_1)^2 e_*$; $E_0 = 0.5\,(F_0 + F_1)^2 e_0$ we obtain $E_*/E_0 = 1 + K_\Delta$. Here,

$$K_\Delta = \frac{4c_0\Delta s}{\pi F_1}\sqrt{1-(F_0/F_1)^2} \qquad (F_1 > F_0) \tag{39}$$

Formula (39) is valid also for torsional vibrations. Here Δs is transformed into the angular displacement in the clearance $\Delta\varphi$, c_0 into the torsional rigidity, F_0 and F_1 and into the torques.

The parameter K_Δ can serve as an effective energy criterion for evaluation of the dynamic effect from the clearance. This criterion is increasing with clearance Δs and stiffness coefficient c_0, and is diminishing with the increase of F_1 and F_0.

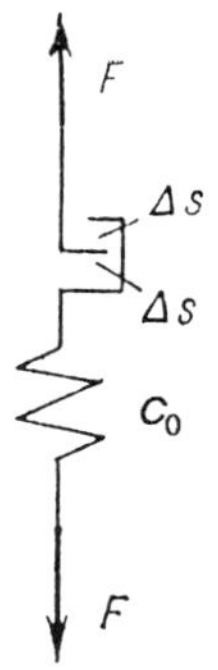

Figure 8. To the analysis of dynamic effect from taking up of the clearance.

At $F_1 \le F_0$ we have $K_\Delta = 0$, which corresponds to the clearance which is fully taken up during the whole kinematic cycle.

Analysis of the experimental results and practical experience show that to limit the level of vibrations caused by clearances, the condition $K_\Delta < 0.1–0.2$ should be satisfied.

Sometimes the dynamic effect of taking up the clearances in the pivots softens somewhat due to a partial conjugate action between the contracting cylindrical surfaces [44, 49]. Different aspects of the problems clearance during vibration analyses of mechanisms are discussed in [2, 7, 23, 27, 28].

CHAPTER
TWO

BRANCHED DRIVES

§7. MODEL WITH ELASTIC CAMSHAFT (GENERAL CASE)

Determination of "natural" frequencies and nonstationary modal coefficients. As a dynamic model of the first level (the "global" model) we consider the torsional vibratory system with distributed parameters corresponding to the camshaft from which n mechanisms are branched. They are presented as subsystems with lumped parameters (Fig. 9).

Suppose one branch corresponds to the drive and the other $n-1$ to the cyclic mechanisms. Each branch has its subcript s; the subscript $j = 1, \ldots, s_*$ indicates the number of the element of the chain s, with counting from the camshaft. The element may be an inertial one (J_{sj}, m_{sj} are the moment of inertia, mass), elastic (c_{sj} is the stiffness coefficient) or serve as a kinematic analog of the mechanism performing the prescribed program transformation of the absolute input coordinate (Π_{sj} is the position function). To simplify indexing and programming, the convention for the notions "input" and "output" for the drive mechanisms is in the direction from the camshaft, i.e., in the same way as for other mechanisms. The subscript s characterizes simultaneously both the cross section of the camshaft in which the branch s is attached, and the segment of the camshaft between the cross sections $s-1$ and s.

Any of coordinates φ which are characterizing the absolute displacement of the arbitrary cross section of the shaft or the displacement of the element sj may be considered as the sum of the "ideal" value corresponding to the programmed motion, ignoring the elasticity of the members (φ^*, φ^*_{sj}) and the dynamic error

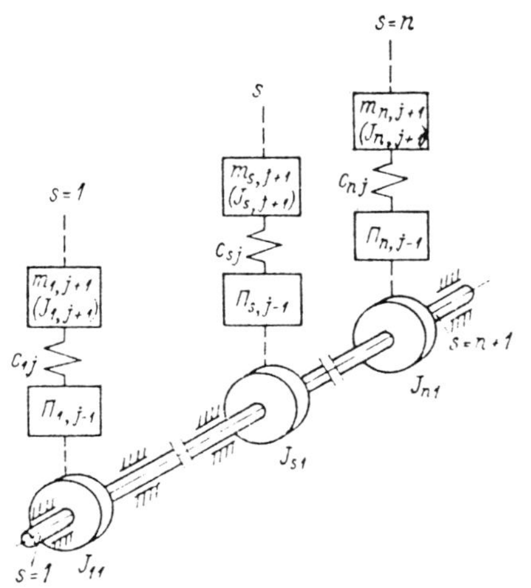

Figure 9. A dynamic model of the branched system of mechanisms with an elastic camshaft.

caused by the vibratory process ($\Delta\varphi$, q_{sj}): $\varphi = \varphi^* + \Delta\varphi$, $\varphi_{sj} = \varphi^*_{sj} + q_{sj}$. For the frequency analysis of the system we assume $\varphi^*_{sj} \equiv 0$, $\varphi^* \equiv 0$, and exclude from the consideration both dissipative and excitation forces. The particular solution of the obtained (for this case set) homogeneous differential equations can be presented as

$$q_{sj} = a_{sj}(\tau)\cos\Phi(t) \qquad \Delta\varphi = X(x, \tau)\cos\Phi(t) \tag{40}$$

where τ is the "slow" time.

At slowly changing parameters, according to relationships (24), we have $\ddot{q}_{ij} \approx -p^2(\tau)\, q_{ij}$ and $\Delta\ddot{\varphi} \approx -p^2(\tau)\, \Delta\varphi\, (x, t)$, where $p(\tau) = d\Phi/dt$ is the changing "natural" frequency. As it has been shown in §5, in this case we may use the apparatus of modified transition matrices.

Let us isolate the cross section s of the shaft, with the initial cross section being taken as zero of the coordinate x_s. The function $X_s(x_s, \tau)$ defining, according to (40), the nonstationary form of vibrations in the segment s, must satisfy the differential equation

$$\partial^2 X_s/\partial x_s^2 + p^2(\tau) X_s/g_s^2 = 0 \tag{41}$$

Here, $g_s = \sqrt{GI_s/\rho_s}$; G is the shear modulus; I_s, ρ_s is the polar moment of inertia and the moment of inertia of mass of the unit length in the segment s; $0 \le x_s \le \Delta l_s$.

The following matrix notation (see Table 1) corresponds to the solution of Eq. (41)

$$\begin{bmatrix} X_s(x_s, \tau) \\ GI_s X'_s(x_s, \tau) \end{bmatrix} = \begin{bmatrix} \cos\vartheta_s & \sigma_s p^{-1}\sin\vartheta_s \\ -\sigma_s^{-1} p \sin\vartheta_s & \cos\vartheta_s \end{bmatrix} \begin{bmatrix} X_s(0, \tau) \\ GI_s X'_s(0, \tau) \end{bmatrix} \tag{42}$$

Here, $\sigma_s = (GI_s \rho_s)^{-0.5}$; $\vartheta_s = px_s/g_s$; $X'_s = \partial X_s/\partial x_s$ (for steel shafts in the SI system $\sigma \approx 4 \cdot 10^{-7} d^{-4}$ kg^{-1} · m^{-2} · s).

Further, consider the branch s, for which we will write the correlation between amplitudes a_{sj} and loads Q_{sj} in the cross section of the camshaft ($j = 0$), and in the end of the branch ($j = s_*$)

$$\begin{bmatrix} a_{ss_*} \\ Q_{ss_*} \end{bmatrix} = \begin{bmatrix} A_s & B_s \\ C_s & D_s \end{bmatrix} \begin{bmatrix} a_{s0} \\ Q_{s0} \end{bmatrix}$$

Here, A_s, B_s, C_s, D_s are the components of the transition matrix Γ_s, which are determined for the series connection by multiplying in the reverse order of the transition matrices of all the elements of the branch (see §5 and Table 1).

It is easy to make sure that $Q_{s0} = -R_s\, a_{s0}$, and if the end of the branch s is fixed ($a_{ss*} = 0$), then $R_s = A_s/B_s$, and at the free end ($Q_{ss*} = 0$) we have $R_s = C_s/D_s$. Thus the function R_s is equal to the momentary dynamic stiffness of the branch s.

Now let's consider the subsystem consisting of the segment s of the camshaft and the branch s. On the junction of the subsystems s and $s + 1$ (the cross section s) considering the adopted sign rule (see §5), we may write the following boundary conditions $X_s(\Delta l_s) = X_{s+1}(0)$; $Q_s = GI_sX_s'(\Delta l_s) = GI_{s+1}X_{s+1}'(0) + Q_{s0}$. These boundary conditions correspond to the recurrent relationships

$$\left.\begin{aligned} K_s &= K_{s-1}\cos\theta_s + N_{s-1}\sin\theta_s \\ N_s &= \beta_s(-K_{s-1}\sin\theta_s + N_{s-1}\cos\theta_s + \sigma_s p^{-1}R_sK_s) \end{aligned}\right\} \tag{43}$$

where $K_s = X_s(0)$; $N_s = \sigma_{s+1}p^{-1}Q_s$; $\theta_s = \theta_s(\Delta l_s)$; $\beta_s = \sigma_{s+1}\sigma_s$ (at $s = n$ it should be assumed that $\beta_s = 1$).

Relationships (43) in the matrix form are

$$\begin{bmatrix} K_s \\ N_s p\sigma_{s+1}^{-1} \end{bmatrix} = \mathbf{V}_s \begin{bmatrix} K_{s-1} \\ N_{s-1}p\sigma_s^{-1} \end{bmatrix}$$

where

$$\mathbf{V}_s = \begin{bmatrix} 1 & 0 \\ R_s & 1 \end{bmatrix} \begin{bmatrix} \cos\theta_s & \sigma_s p^{-1}\sin\theta_s \\ -\sigma_s^{-1}p\sin\theta_s & \cos\theta_s \end{bmatrix}$$

If by means of the transition matrices of the separate subsystems $\mathbf{V}_s$ all n segments were passed step by step, and considering that the load amplitudes in the initial cross section are equal to zero, then

$$\begin{bmatrix} K_n \\ 0 \end{bmatrix} = \begin{bmatrix} v_{11}(p) & v_{12}(p) \\ v_{21}(p) & v_{22}(p) \end{bmatrix} \begin{bmatrix} 1 \\ 0 \end{bmatrix} \tag{44}$$

Here $X_1(0) = 1$; $K_n = X_{n-1}(\Delta l_{n-1})$; $v_{11}, v_{12}, v_{21}, v_{22}$ are the elements of the transition matrix $\mathbf{V} = \prod_{s=n}^{1} \mathbf{V}_s$.

It follows from the matrix equation (44) that $v_{21}(p) = 0$; this relationship serves as a formal frequency equation. The other form of this equation can be obtained by means of the recurrent relationships (43) if we assume in these relationships $K_0 = 1, N_0 = 0, N_n = 0$. The latter condition is corresponded by

$$U(p) = -K_{n-1}\sin\theta_n + N_{n-1}\cos\theta_n + \sigma_n p^{-1} R_n K_n = 0 \quad (45)$$

If in the n-th cross section a mechanism is absent, then in Eq. (45) it should be assumed $R_n = 0$; then $N_{n-1} = K_{n-1}\operatorname{tg}\theta_n$.

Here, the formal frequency equation is transcendental and has an infinite number of roots p_r ("natural" frequencies). For analysis of the influence of the vibratory system's parameters, it may be more convenient to break the solution of Eq. (45) into two steps. Let, for example, the drive mechanism correspond to the branch n. Then on the basis of (45), the formal frequency equation can be written as

$$\Phi(p) = H_m(p) - H_d(p) = 0 \quad (46)$$

where $H_m(p) = (-K_{n-1}\sin\theta_n + N_{n-1}\cos\theta_n)/K_n$; $H_d(p) = -\sigma_n R_n/p$.

Here and below, the subscript m at the function H corresponds to the machine, and the subscript d corresponds to the drive. If the functions $H_m(p)$ and $H_d(p)$ are represented as plots, the points of the intersection of these plots will correspond to the "natural" frequencies p_r.

The procedure for determining the reactive torque is more complicated if the drive mechanism is located in the intermediate cross section of the shaft $s = k$. In this case, for the fixed value p the function $U(p)$ should be determined twice by Eq. (45), namely: $U = U_1 \neq 0$ at $K_0^{(1)} = 1$, $R_k^{(1)} = 0$ and $U = U_2 \neq 0$ $K_0^{(2)} = 0$; $\sigma_k R_k^{(2)}/p = 1$. After this we have $H_m = U_1(p)/(U_2(p).K_k)$. The function $H_d(p)$ is determined, as before, by changing the subscript n into k.

Next, the non-stationary modal coefficients are determined. The values $K_s^{(r)} = K_s(p_r)$, acting as modal coefficients for the corresponding cross sections of the shaft, are determined directly from the recurrent relationships (43), or by means of the transition matrices at $p = p_r$.

In the further discussion, the ensemble of the amplitudes on the boundaries of the segments $K_0, K_1, ..., K_n$ will be called the *stroboscopic mode of vibrations*. The basis for such a name is a certain analogy to the stroboscopic effect, since in this case out of all the ensemble $X(x, \tau)$, the discrete values corresponding to certain cross sections are "illuminated". By analogy we may call the sequence of values $Q_s = pN_s/\sigma_s$ *the stroboscopic load mode.*

Inside each segment the mode is described by relationships

$$X_s^{(r)}(x_s, \tau) = K_{s-1}^{(r)}\cos\vartheta_s^{(r)} + N_{s-1}^{(r)}\sin\vartheta_s^{(r)} \quad (47)$$

For the element sj the modal coefficient $a_{sj}^{(r)}$ is determined by the formula

$$a_{sj}^{(r)} = K_s^{(r)}\alpha_{sj}^{(r)} = [A_{sj}^*(p_r) - B_{sj}^*(p_r)R_s(p_r)]K_s^{(r)} \quad (48)$$

where A_{sj} and B_{sj}^* are corresponding elements of the first row of the matrix $\Gamma_{sj} = \prod_{\nu=j}^{1}\Gamma_{s\nu}$, with the matrix $\Gamma_{s\nu}$ coinciding with the matrix Γ_{sj}.

Realization of computation procedure on computer. When calculating on a computer, the initial information about the drive and mechanisms is given by the matrices Γ_{sj} with the elements $A_{sj}, B_{sj}, C_{sj}, D_{sj}$. The sequence of these matrices by j must strictly correspond to the sequence of the elements in the given branch when advancing from the camshaft. Since at the inertial element the coefficient $C_{sj} = -p^2 J_{sj}$ is a changing value when the frequency p is varied, we can put the value $C^*_{sj} = J_{sj}$ in the initial inertial matrix so that C_{sj} is generated in the computational process as $C_{sj} = -p^2 C^*_{sj}$. Here, the distortion does not occur in the corresponding elements of the noninertial matrices since in them $C_{sj} = C^*_{sj} = 0$. When forming the elements of the matrices representing the kinematic analog Π_{sj}, in each separate case a subprogram is required realizing the dependence of the position function on the "ideal" input coordinate. The necessity for a separate subprogram may also arise in such cases when the reduced stiffness or the reduced moment of inertia of any element depend on the angular position of the main shaft.

For each branch s the format of boundary conditions at the end must be stated (for example, $v_s = 1$ for the free end, v_s for the fixed end), and for the camshaft the parameters σ_s and g_s must be calculated in each segment.

The computations are run for $p \in [p_{\mathrm{I}}, p_{\mathrm{II}}]$, with the boundary values $p_{\mathrm{I}}, p_{\mathrm{II}}$ being determined from engineering considerations. The given range of values p at the computation should be passed with a certain stepsize, with formation of the matrix Γ_s taking place at each step (see §7). After this, the functions K_s and N_s are determined by recurrent relationships (43) (if, $K_0 = 1$, then $N_0 = 0$) and the function $U = N_n$ is determined by formula (45).

When the function U changes its sign, the root p is sought with an accuracy determined by the restrictions $|U| < |U|$. For the value p obtained in such a way from (47) and (48), the calculation of the nonstationary modal coefficients is performed, followed search for the next root.

At the two-step calculation of frequencies (see above) for each value p the dimensionless functions H_{m} and H_{d} are printed out. This algorithm can be realized also on small computers.

Determination of "natural" frequencies with absolutely rigid camshaft. In this case the formal frequency equation takes a simple form [11]

$$\sum_{s=1}^{n} R_s(p) = 0 \tag{49}$$

Suppose, for example, that the drive mechanism $s = 1$ is represented as a series: $J_{11} - \Pi_{12} - c_{13}$—fixation, and each of the branches $s = 2, ..., n$ as a circuit $\Pi_{s1} - c_{s2} - J_{s3}$ (counting of elements in all the circuits is adopted in the direction from the camshaft). Then the transition matrices for the branches are determined as $\Gamma_s = \Gamma_{s3}\,\Gamma_{s2}\,\Gamma_{s1}$. Using the formulae of Table 1 we have $A_1 = \Pi'_{12} - J_{11}p^2 /(c_{13}\Pi'_{12})$; $B_1 = (c_{13}\Pi'_{12})^{-1}$; at $s \geq 2$ $A_s = \Pi'_{s1}$; $B_s = (c_{s2}\Pi'_{s1})^{-1}$; $C_s = -p^2 J_{s3}\Pi'_{s1}$; $D_s = (\Pi'_{s1})^{-1}(1 - p^2 J_{s3} c_{s2}^{-1})$, and the formal equation (49) becomes

$$A_1/B_1 + \sum_{s=2}^{n} C_s/D_s = 0 \tag{50}$$

The next case is one of equal *partial* frequencies; by this we mean the "natural" frequency of the branch when its input cross section located on the camshaft is fixed. For the considered example, the partial frequency at $s \geq 2$ is equal to $p_{s*} = \sqrt{c_{s2}/J_{s3}}$. If at $s = 2, \ldots, n$ all the branches are identical ($C_s = C$; $D_s = D$), then Eq. (50) breaks down into two independent equations and $D^{n-2}(p) = 0$ and $A_1(p)/B_1(p) + (n-1)\,C(p)/D(p) = 0$ $(D \neq 0)$. The roots of the first equation are the "natural" frequencies coinciding with the partial frequencies (in this sense). The multiplicity of this group of roots being by one less than the number of the identical mechanisms. The second equation corresponds to such a drive in which $n - 1$ identical mechanisms are identical to one mechanism at $R = (n-1)\,C/D$.

In many drives of the machines and the automatic transfer lines cases are possible when from the common camshaft the mechanisms which form a number of repeating identical blocks branch off, the mechanisms being different within each block. Suppose that the drive consists of the driving mechanism ($s = 1$) and $n - 1$ mechanisms united into k similar blocks, each block including N different mechanisms ($n - 1 = k \times N$). Then Eq. (50) breaks down into two groups of equations:

$$\left.\begin{aligned} &D_j^{k-1}(p)=0 && (j=1,\ \ldots,\ N) \\ &A_1(p)/B_1(p)+k\sum_{j=1}^{N} C_j(p)/D_j(p)=0 && (D_j \neq 0) \end{aligned}\right\} \qquad (51)$$

Retaining the structure of the dynamic model considered in the previous example, let's assume now that twelve mechanisms make up three identical blocks by four different mechanisms in each block ($k = 3$, $N = 4$). After substituting into Eqs. (51) the above cited specific values of elements of the transition matrices we obtain:

$$(\Pi'_j)^{-1}(1-p^2J_j/c_j)^2=0 \qquad (j=1,\ \ldots,\ 4)$$

$$c_{13}(\Pi'_{12})^2-J_{11}p^2-3p^2\sum_{j=1}^{4}(\Pi'_j)^2J_j(1-p^2J_j/c_j)^{-1}=0$$

The first group of the equations has eight roots; the second group has five roots.

Specifics of frequency spectrum of drives with elastic camshaft, including identic mechanisms. Consider now the drive mechanism ($s = 1$) with the elastic distributing shaft (camshaft), to which twelve cyclic mechanisms are attached forming four groups of the similar mechanisms. In the same way as for the analogous drive with the absolutely rigid shaft (see above), each of the vibratory circuits corresponding to the drive and the mechanisms has a single degree of freedom. However, the shaft, in accordance with the adopted model, now represents a torsional system with distributed parameters. This predetermines the infinite number of degrees of freedom in the considered system (in contrast to 13 degrees of freedom in the previous example).

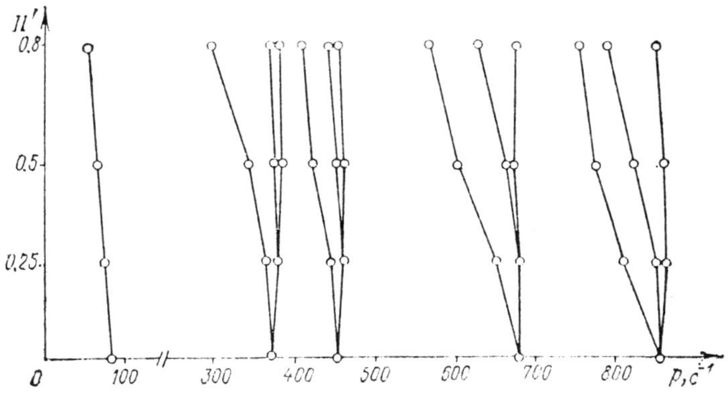

Figure 10. The plots of dependences of "natural" frequencies on the values of the first transmission function.

The mechanisms are presented as series connections of the following discrete elements (counting proceeds from the main shaft): the driving mechanism – J_{11} – Π_{12} – J_{13} – c_{14}-clamping; actuators –Π_{s1} – c_{s2} – J_{s3} (s = 2, ..., 13). The identical mechanisms correspond to the following values of s: group 1 to s = 2, 7, 13; group 2 to s = 3, 6,12; group 3 to s = 4, 9, 11; group 4 to s = 5, 8, 10.

Fig. 10 gives a characteristic set of plots for this example, representing the connection between the "natural" frequencies p and the first transmission function Π'; the calculated points are shown as circles. On the abscissa, the partial frequencies corresponding to $\Pi' = 0$ are arranged. Their lower value corresponds to the partial frequency of the driving mechanism, while the other four correspond to the multiple frequencies of various actuators (see above). When Π' deviates from zero, the frequencies split into three branches. Of interest is the rigorous succession of nonstationary modes of vibrations, which is characteristic for each bundle of frequencies. Namely, the first branch corresponds to a single-node mode, the second one to a two-node mode, and the third one to a special type of the exponential mode with which an interesting effect of spatial damping is associated (more details in §§9, 10, 12). The noted recurrence of the vibratory modes, which is observable on the main shaft at different values of P_r, does not mean, of course, the identity of modes in the whole system since the modes in the branches are substantially different.

Transition to the "local" dynamic model. The local dynamic model is described by the differential equation (33). First let's determine the inertial coefficient J_r, for which a cross section should be previously chosen to which all the elements of the system vibrating in the considered mode r are reduced. Suppose this cross section is located on the branch $s = s^*$ and has the ordinal number $j = j^*$. Then, using the condition of balance for kinetic energy for this mode, we have

$$J_r(K_r^*\alpha_r^*)^2 = \sum_{S=1}^{n} \left(\rho_s \int_0^{\Delta l_s} X_s^2 dx + \sum_{j=1}^{j_{\max}} J_{sj} K_s^2 \alpha_{sj}^2 \right)_{p=p_r} \qquad (52)$$

Here, K_r^* and α_r^* correspond to the cross section s^* and j^* (if the chosen cross section coincides with the cross section s of the camshaft, then $K_r^* = K_s$ and $\alpha_r^* = 1$).

After the substitution of relationship (47) into (52) and integration, obtain

$$J_r = 0.5\,(K_r^*\alpha_r^*)^{-2} \sum_{s=1}^{n} \rho_s [(K_{s-1}^2 + N_{s-1}^2)\Delta l_s + 0.5 g_s p^{-1}(K_{s-1}^2 - N_{s-1}^2)\sin 2\theta_s + g_s p^{-1} K_{s-1} N_{s-1}(1-\cos 2\theta_s)]_{p=p_r} + \sum_{s=1}^{n} \sum_{j=1}^{j_{\max}} J_{sj}(\zeta_{sj}^r)^2 \tag{53}$$

where $\zeta_{sj}^r = [K_s\alpha_{sj}/(K_r^*\alpha_r^*)]_{p=p_r}$.

The function ζ_{sj}^r, which is the normalized value of the modal coefficient, characterizes coupling in the system. As the analysis [14] has shown, relatively larger values of ζ_{sj}^r correspond to the mechanisms with close partial frequencies, and thus are characterized by stronger coupling. The choice of the cross section s^*j^*, in order to avoid the degeneracy of the local model, should be carried out in such a way that the possibility of $\zeta_{sj}^r \to \infty$ should be excluded at any t. Practically, it is achieved by choosing a cross section on the branch whose partial frequency is close to the "natural" frequency p_r.

The coefficient of dissipation ψ_r, reduced to the mode r, is determined as

$$\psi_r = V_r^{-1} \sum_{s=1}^{n} (\psi_{0s} V_{0s} + \sum_{j=1}^{j_{\max}} \psi_{sj} V_{sj})_{p=p_r} \tag{54}$$

Here, ψ_{0s}, ψ_{sj} are the coefficients of dissipation of the segment s of the camshaft and of the element sj; V_{0s}, V_{sj} is the potential energy in the mode r corresponding to these elements. Here,

$$V_{0s} = 0.5 G I_s (K_r^*\alpha_r^*)^{-2} \int_0^{\Delta l_s} (dX_s/dx_s)^2 dx$$
$$= 0.5 p \sigma_s^{-1} (K_r^*\alpha_r^*)^{-2} [(K_{s-1}^2 + N_{s-1}^2)\theta_s + 0.5\sin 2\theta_s (K_{s-1}^2 - N_{s-1}^2) + K_{s-1}N_{s-1}(1-\cos 2\theta_s)]_{p=p_r}$$

$$V_{sj} = 0.5(\zeta_{s,\,j+1}^r - \zeta_{s,\,j-1}^r)^2 B_{sj}^{-1} \qquad (B_{sj} \neq 0)$$

$$V_r = \sum_{s=1}^{n} (V_{0s} + \sum_{j=1}^{j_{\max}} V_{sj}) = p_r^2 J_r$$

(at $B_{sj} = 0$ it should be assumed that $V_{sj} = 0$).

If in all the elements the coefficient of dissipation may be assumed to be the same ($\psi_{0s} = \psi_{sj} = \psi^*$) then from (54) we have $\psi_r = \psi^*$.

This result in the first approximation can be used when there is no reliable information on the distribution of the dissipative factors between individual elements of the system.

The right-hand side of the differential equation (33), corresponding to the external excitation reduced to the mode r, is determined as

$$W_r = (J_r \alpha_r^* K_r^*)^{-1} \sum_{s=1}^{n} \int_0^{\Delta k_s} [\mu_s(x_s, t) - \rho_s \ddot{\varphi}^0(t)] X_s^{(r)}(x_s, \tau) dx + \sum_{s=1}^{n} \sum_{j=1}^{j_{\max}} (F_{sj} - J_{sj} \ddot{\varphi}^0_{sj}) \zeta^r_{sj} \tag{55}$$

Here, F_{sj} is the torque or force applied to the element sj; φ^0_{sj}, φ_0 are the absolute coordinates of the element sj and the camshaft in the programmed motion; μ is the distributed torque applied along segment s of the camshaft.

If in the programmed motion the shaft is rotating uniformly, then in (55) it should be assumed $\ddot{\varphi}^0 = 0$.

The functions η_r obtained during solution of the differential equation (33) are distributed between the coordinates of the system in accordance with the relationship $q_{sj} = \sum_{r=1}^{r_{\max}} \eta_r \zeta^r_{sj}(\tau)$.

§8. STRUCTURAL TRANSFORMATIONS OF MODELS WITH INTERMEDIATE BRANCHES AND CONSTRUCTION OF GENERALIZED MODEL

The kinematic circuits of many machine drives have "tree" structures; here the main carrying subsystem ("trunk") consists of a number of blocks connected in series and having branches which in turn may have their branches and so on (Fig. 11a). In cases when the main carrying subsystem is not structurally clearly

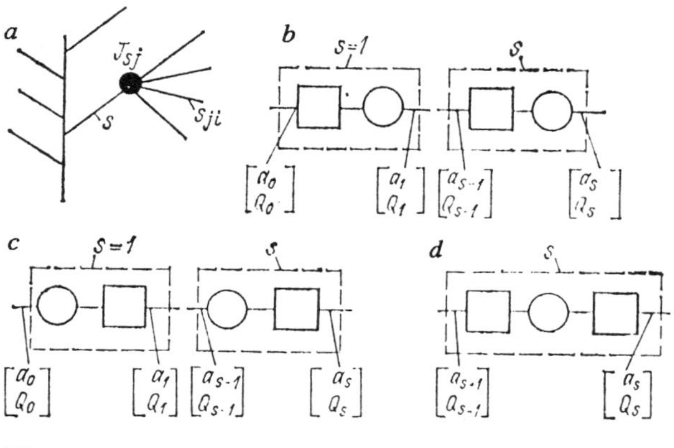

Figure 11. Structural transformations of the dynamic model.

expressed, it is convenient to choose as a "trunk" of the model either the most loaded chain of the drive or the longest branch. By means of the transition matrix apparatus considered above we can carry out structural transformations simplifying the dynamic analysis of the system. Suppose that a number of additional vibratory chains branch off from some inertial element J_{sj}. As it has been shown in §7, amplitude of reactive torque (or force) acting from the additional branch i on the element J_{sj} is equal to $\Delta Q_{sji} = -R_{sjt}\ a_{sj}$, and the total reactive component is accordingly equal to

$$\Delta Q_{sj} = -a_{sj} \sum_{i=1}^{n_{sj}} R_{sji}$$

Here, $R_{sji} = C_{sji}/D_{sji}$ or $R_{sji} = A_{sji}/B_{sji}$ at the free and fixed ends of the branch sji, respectively; $A_{sji}, B_{sji}, C_{sji}, D_{sji}$ are the elements of the transition matrix of the branch sji (see §7).

The transition through the element sj, subject to the sign rule, corresponds to the following matrix:

$$\begin{bmatrix} a_{sj} \\ Q_{sj} \end{bmatrix} = \begin{bmatrix} 1 & 0 \\ -p^2 J_{sj} + R_{sj} & 1 \end{bmatrix} \begin{bmatrix} a_{s,\ j-1} \\ Q_{s,\ j-1} \end{bmatrix} \tag{56}$$

where

$$R_{sj} = \sum_{i=1}^{n_{sj}} R_{sji}$$

Thus, from (56) we may make a conclusion that the dynamic effect from the side chains is equivalent to the introduction of some fictitious additional moment of inertia (or mass) $\Delta J_{sj} = -R_{sj}/p^2$, and the inertial coefficient corrected considering all branches becomes $J'_{sj} = J_{sj} + \Delta J_{sj}(p)$.

By means of similar successive transitions, the model presented in Fig. 11*a* takes the form shown in Fig. 11*b*; here, the subsystems of the "trunk" proper are designated by squares and those substituting the branches are designated by circles. Each pair of such subsystems forms a block whose subscript corresponds with the subscript of the branch. As before, the sequence of amplitude functions on the boundaries of blocks $a_0, a_1, ..., a_n$ would characterize the "stroboscopic" mode of vibrations, while the sequence of the load amplitudes $Q_0, Q_1, ..., Q_n$ would characterize the "stroboscopic" mode of loads. For an arbitrary value $s = k$ ($s = 1, ..., n$) we have

$$\begin{bmatrix} a_k \\ Q_k \end{bmatrix} = \prod_{s=k}^{1} \Gamma_s \begin{bmatrix} a_0 \\ Q_0 \end{bmatrix} \tag{57}$$

where for the first model in Fig. 11*b*

$$\Gamma_s = \begin{bmatrix} 1 & 0 \\ R_s & 1 \end{bmatrix} \begin{bmatrix} A_{0s} & B_{0s} \\ C_{0s} & D_{0s} \end{bmatrix} = \begin{bmatrix} A_{0s} & B_{0s} \\ R_s A_{0s} + C_{0s} & R_s B_{0s} + D_{0s} \end{bmatrix} \tag{58}$$

The subscript $0s$ corresponds to the elements of the transition matrix of the segment "trunk", which is a part of block s.

For the second model (Fig. 11c)

$$\mathbf{\Gamma}_s = \begin{bmatrix} A_{0s} & B_{0s} \\ C_{0s} & D_{0s} \end{bmatrix} \begin{bmatrix} 1 & 0 \\ R_s & 1 \end{bmatrix} = \begin{bmatrix} A_{0s}+B_{0s}R_s & B_{0s} \\ C_{0s}+D_{0s}R_s & D_{0s} \end{bmatrix} \tag{59}$$

In Fig. 11d is shown the block of a general type in which branches are located between two segments of the trunk s' and s''. Here

$$\mathbf{\Gamma}_s = \begin{bmatrix} (A''_{0s}+B''_{0s}R_s)A'_{0s}+B''_{0s}C'_{0s} & (A''_{0s}+B''_{0s}R_s)B'_{0s}+B''_{0s}D'_{0s} \\ (C''_{0s}+D''_{0s}R_s)A'_{0s}+D''_{0s}C'_{0s} & (C''_{0s}+D''_{0s}R_s)B'_{0s}+D''_{0s}D'_{0s} \end{bmatrix} \tag{60}$$

where one prime corresponds to the element s', and two primes correspond to the element s''.

Relationships (58) and (59) can be obtained from (60) when $A''_{0s} = D''_{0s} = 1$; $B''_{0s} = C''_{0s} = 0$ and $A'_{0s} = D'_{0s} = 1$; $B'_{0s} = C'_{0s} = 0$, respectively.

With the proposed approach, the element of the trunk or any branch can have a discrete or distributed dynamic structure which is reflected in the corresponding transition matrix.

Since the system became a series connection of subsystems after the performed structural transformations, a further analysis does not differ from that presented above. In particular, if the system has free ends and consists of n blocks, then the formal frequency equation can be obtained from the boundary conditions $Q_0 = 0$; $Q_n = 0$ at the arbitrary assignment of the amplitude in one of the cross sections (for example, $a_0 = 1$).

§9. USE OF PROPERTIES OF REGULAR SYSTEMS FOR ANALYSIS OF BRANCHED DRIVES WITH IDENTICAL MECHANISMS

Preliminary remarks. Identical cyclic mechanisms are widely used in the machines and the transfer lines in order to perform repetitive technological and transport operations. In such cases, there are a number of mechanisms, identical in their structural and kinematic characteristics, which are connected by a common drive and by a distribution shaft (camshaft). As it has been shown in §7, with an absolutely rigid camshaft and n identical mechanisms the multiple frequencies appear with the degree of multiplicity $n-1$. With an elastic camshaft, application of general analytical methods presented in §7 and of corresponding computational algorithms has shown that in the vicinity of the partial frequencies of the mechanisms the spectrum of the "natural" frequencies was quite dense. It often causes omission of some frequencies and the poor conditioning of the computational procedure. On the other hand, repetitiveness of the subsystems in the vibratory chain (*regularity*) allows for application of some special analytical procedures. They result in presenting the formal frequency equation and modes of vibrations subject to the particular boundary conditions in the analytical form [4, 22].

Frequency and modal analysis of regular systems by means of apparatus of finite differential equations. Retaining unchanged the initial premises, the form of solution, and the general procedure presented in §7, we will refer directly to the generalized model (see Fig. 11*b*), for which the stroboscopic modes are described by the matrix equality (57). The recurrent relationships corresponding to (57) and (58) when passing through the block *s* are

$$\left.\begin{aligned} a_s &= A_{0s}a_{s-1} + B_{0s}Q_{s-1} \\ Q_s &= (R_sA_{0s} + C_{0s})a_{s-1} + (D_{0s} + R_sB_{0s})Q_{s-1} \end{aligned}\right\} \tag{61}$$

The recurrent relationships (61) represent the *homogeneous system of finite differential equations*, whose solution we seek in the form $a_s = a_{s-1}\lambda_s$, $Q_s = Q_{s-1}\lambda_s$. Then

$$\left.\begin{aligned} &(A_{0s} - \lambda_s)a_{s-1} + B_{0s}Q_{s-1} = 0 \\ &(C_{0s} + R_sA_{0s})a_{s-1} + (D_{0s} + R_sB_{0s} - \lambda_s)Q_{s-1} = 0 \end{aligned}\right\}$$

Ignoring the trivial zero solution of this system, we would require

$$\begin{vmatrix} A_{0s} - \lambda_s & B_{0s} \\ C_{0s} + R_sA_{0s} & D_{0s} + R_sB_{0s} - \lambda_s \end{vmatrix} = 0 \tag{62}$$

The root of Eq. (62) is the eigenvalue of the transition matrix Γ_s, with

$$\lambda_s = \varkappa_s \pm \sqrt{\varkappa_s^2 - 1} \tag{63}$$

where

$$\varkappa_s = 0.5(A_{0s} + D_{0s} + R_sB_{0s}) \tag{64}$$

Here it is taken into account that det $\Gamma_s = 1$.

Thus, to determine the eigenvalue λ_s it is necessary to know only the trace of the transition matrix $\mathrm{Sp}\Gamma_s = 2\kappa_s$. Using relationship (60) for the block of the general form, we may show that κ_s and λ_s do not depend on the place of the location of the branch within the block.

Depending on the value κ_s, there may be three cases.

Case 1. Suppose $|\kappa_s \leq 1|$. Then Im $\lambda_s \neq 0$, and the characteristic numbers λ_s are mutually conjugated complex numbers whose magnitude is one. Here, $\lambda_s = \cos\gamma_s \pm i\sin\gamma_s = e^{\pm i\gamma_s}$, where $\cos\gamma_s = \kappa_s$; $\sin\gamma_s = \sqrt{1-\kappa_s^2}$. The solution of the differential equation has the form $a_s = C_1^{(s-1)}e^{i\gamma_s} + C_2^{(s-1)}e^{-i\gamma_s}$, where $C_1^{(s-1)}$, $C_2^{(s-1)}$ are some complex numbers.

Suppose that at $s_1 \leq s \leq s_2$, all the blocks are identical, hence $\lambda_s = \lambda, \gamma_s = \gamma$ Then

$$a_s = C_1\exp[i(s - s_1 + 1)\gamma] + C_2\exp[-i(s - s_1 + 1)\gamma]$$

By the known procedure this expression is reduced to the trigonometric form

$$a_s = h_1\cos[(s-s_1+1)\gamma] + h_2\sin[(s-s_1+1)\gamma] \tag{65}$$

where h_1, h_2 are real arbitrary constants.

In order to determine Q_s we may use one of the following relationships:

$$Q_s = (a_{s+1} - A_0 a_s)B_0^{-1} \qquad (s_1 \leq s \leq s_2 - 1) \tag{66}$$

$$Q_s = [(D_0 + RB_0)a_s - a_{s-1}]B_0^{-1} \qquad (s_1 - 1 \leq s \leq s_2) \tag{67}$$

(the subscript s at the identical blocks is omitted).

Expressions (65) and (66) or (67), which determine the stroboscopic modes by amplitudes and loads, depend on the yet unknown value of the "natural" frequency p_r, which can be determined subject to the particular boundary conditions (see below).

Case 2. Suppose $\kappa > 1$ ($\mathrm{Im}\,\lambda_s = 0$). Assuming $\mathrm{ch}\gamma = \kappa_s$ and carrying out similar transformations, we obtain expressions for a_s and Q_s differing from the corresponding expressions (65), (66), and (67) only in the fact that the trigonometric functions are substituted by the hyperbolic function of the same name.

Case 3. Suppose $\kappa_s \leq -1$. The condition $\mathrm{ch}\gamma = \kappa_s$ is now safisfied only at $\gamma = \gamma^0 + i\pi$ ($i = \sqrt{-1}$). Here we have $\mathrm{ch}\,(\gamma^0 + i\pi) = \mathrm{ch}\gamma^0\mathrm{ch}i\pi + \mathrm{sh}\gamma^0\mathrm{sh}i\pi = \mathrm{ch}\gamma^0\cos\pi - i\mathrm{sh}\gamma^0\sin\pi = -\mathrm{ch}\gamma^0 \leq -1$. Thus, in this case it should be assumed that $\mathrm{ch}\gamma^0 = -\kappa_s$, with γ^0 now being a real number. On the basis of Moivre's formula it is easly to show that $\mathrm{ch}\,k\gamma = (-1)^k\,\mathrm{ch}k\gamma^0$ and $\mathrm{sh}k\gamma = (-1)^k\mathrm{sh}k\gamma^0$. Subject to these corrections, for this case the formulae obtained for case 2 may be used.

Thus, side by side with the traditional case when the modes of torsional vibrations have a usual trigonometric format, the stroboscopic forms at certain conditions can be described by hyperbolic functions, and in Case 3 the sign of the amplitude function is changing at the transition through each block.

§10. REGULAR MODELS OF BRANCHED STRUCTURE DRIVES

The dynamic model of the drive, consisting of the main shaft which is schematized as a torsional subsystem with distributed parameters and to which the driving mechanism ($s = 1$) and k identical actuators ($s = 2, ..., n = k + 1$) are attached, is presented in Fig. 12. Each mechanism is represented as a series chain of the discrete elements—elastic (c), inertial (J), and kinematic (Π). The sequence of connection of the blocks marked in Fig. 12 by the dashed line corresponds to the generalized model in Fig. 11b. The transition matrix describing the inertial and elastic properties of the shaft in the segment s includes the following elements (see Table 1): $A_{0s} = \cos\theta_s$; $B_{0s} = \sigma_s p^{-1}\sin\theta_s$; $C_{0s} = -\sigma_s^{-1} p\sin\theta_s$; $D_{0s} = \cos\theta_s$; where $\theta_s = p\Delta l_s/g_s$; $g_s = (GI_s/\rho_s)^{0.5}$; $\sigma_s = (GI_s\,\rho_s)^{-0.5}$ (see nomenclature in §5). Since the shaft diameter and the lengths of all segments are the same, the values of the elements of this matrix at $s = 2, ..., n$ do not depend on s. In order to increase the sphere of applicability of this model, it includes two segments of the

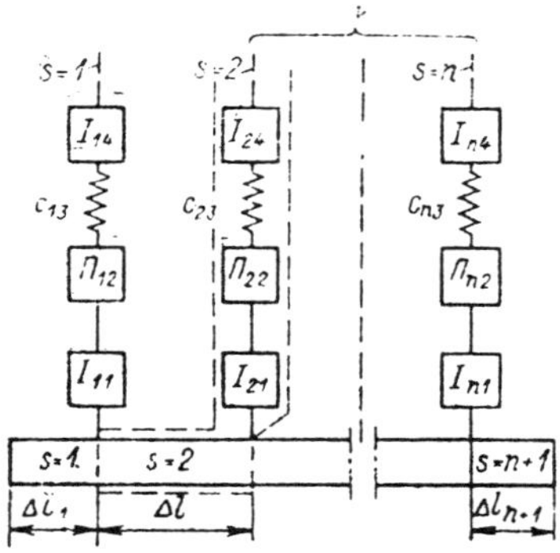

Figure 12. The regular dynamic model of the branched system.

main shaft $s = 1$, $s = n + 1$ whose length may not coincide with the length of the segments of the regular system.

For this model we shall use the general solution presented in the previous section.

The eigenvalue of the transition matrix λ_s depends on the parameter κ_s, which can be determined from Eq. (64) after substitution of the values A_{0s}, B_{0s}, D_{s0} given above

$$\varkappa_s = \cos\theta_s + 0.5 R_s \sigma_s p^{-1} \sin\theta_s \tag{68}$$

As it has been determined above, depending on the value of κ_s, one of three cases takes place in the considered frequency range: $|\kappa| < 1$, $\kappa \geq 1$, and $\kappa \leq -1$. Here and below, for the regular part of the system the subscript s at the same parameters is omitted.

For this model the regular part of the system is limited by the values $s_1 = 2$ and $s_2 = k + 1$, where k is the number of the identical blocks. The boundary conditions for the shaft have the form $Q_0 = 0$ and Q_{n+1} (the ends are free). Then, assuming $a_0 = 1$ on the basis of the recurrent relationships (61), we obtain $Q_1 = R_1 \cos\theta_1 - p^{\sigma-1} \sin\theta_1$, $a_1 = \cos\theta_1$. These values serve as boundary conditions from the left for the regular part of the system. After simple manipulations we obtain a formal frequency equation as $\Phi(p) = H_d(p) - H_m(p) = 0$, where $H_d = (R_1 - p\sigma^{-1} \operatorname{tg}\theta_1)/\Delta c_0$; $\Delta c_0 = (GI)/\Delta l$ is the torsional stiffness coefficient of the shaft segment s at $s = 2, \ldots, n$.

It should be remembered that the function $R_1(p)$, which corresponds to the momentary dynamic stiffness of the driving mechanism, is determined as $R_1(p) = A_1(p)/B_1(p)$ at the fixed end of the chain $s = 1$, and $R_1(p) = C_1(p)/D_1(p)$ at the free end. The function H_m, as well as relations defining the stroboscopic mode of vibrations $a_s = K_s$ and the stroboscopic form of loads $N_s = Q_s \sigma p^{-1}$, are presented in Table 2 (the subscript d at the function indicates driving mechanism, and the subscript m indicates machine). Within each segment s $(0 < x_s < \Delta l)$ the mode of vibrations r is determined by relationship (47), into which $K_{s-1}^{(r)}$ and $N_{s-1}^{(r)}$ corresponding to the considered "natural" frequency p_r should be substituted.

The roots of the equation $R^{-1}(p) = 0$, which are equal to p_*, are points of thickening in the vicinity of which the frequency ranges with the increased density of distribution of the "natural" frequencies are located.

A more detailed analysis of this zone, as well as specifics of the cases discussed above, will be given in the following example.

It is of interest that the value κ (and, consequently, the eigenvalue of the transition matrix) does not depend on the place where the driving member of the mechanism is located in the segment Δl; this factor is reflected only in the stroboscopic modes.

Example. We shall illustrate the computational algorithm and the analysis of the frequency and modal characteristics on an example of a drive in which the driving mechanism ($s = 1$) is schematized as the following sequence of simple elements: J_{11}-Π_{12}-c_{13}-clamp (counted from the main shaft). The identical actuators ($s = 2, ..., k + 1$) are schematized as Π_{s1}-c_{s2}-J_{s3}. Here we assume the following initial data: $J_{11} = 0.208$; $J_{s3} = 0.744\ /k$ kg $\cdot$ m^2; $\Delta c_0 = 0.847 \cdot 10^4\ k$; $c_{s2} = 1.56 \cdot 10^5\ /k$; $c_{13} = 6 \cdot 10^3 N \cdot$ m; $k = 2, 4, 6, 12$ is the number of identical mechanisms; $\Delta l_1 = \Delta l_{n+1} = 0$; $l = 2.408$ m; $\Delta l = l/k$; $\sigma = 0.228$ m$^{-2} \cdot$ kg$^{-1} \cdot$ s; $g = 4.65 \cdot 10^3$ m $\cdot$ s^{-1}; $\Pi'_{12} = 05$; $\Pi'_{s1} \in [0, 0.8]$.

Table 2. Relationships for determining frequency and modal characteristics of regular branched drives

Case	$\varkappa$, γ, γ^0	ξ
1	$\lvert\varkappa\rvert < 1$; $\gamma = \arccos\varkappa$	$\dfrac{\cos(k+1)\gamma - (\cos\theta + \mathrm{tg}\,\theta_{n+1}\sin\theta)\cos k\gamma}{\sin(k+1)\gamma - (\cos\theta + \mathrm{tg}\,\theta_{n+1}\sin\theta)\sin k\gamma}$
2	$\varkappa > 1$; $\gamma = \mathrm{Arch}\,\varkappa$	$\dfrac{\mathrm{ch}(k+1)\gamma - (\cos\theta + \mathrm{tg}\,\theta_{n+1}\sin\theta)\mathrm{ch}\,k\gamma}{\mathrm{sh}(k+1)\gamma - (\cos\theta + \mathrm{tg}\,\theta_{n+1}\sin\theta)\mathrm{sh}\,k\gamma}$
3	$\varkappa < -1$; $\gamma^0 = \mathrm{Arch}\,\lvert\varkappa\rvert$	$-\dfrac{\mathrm{ch}(k+1)\gamma + (\cos\theta + \mathrm{tg}\,\theta_{n+1}\sin\theta)\mathrm{ch}\,k\gamma}{\mathrm{sh}(k+1)\gamma + (\cos\theta + \mathrm{tg}\,\theta_{n+1}\sin\theta)\mathrm{sh}\,k\gamma}$

Case	H_m	$K_s^{(r)}$ $(p = p_r)$
1	$(\cos\gamma - \cos\theta - \xi\sin\gamma)\theta/\sin\theta$	$[\cos(s-1)\gamma - \xi\sin(s-1)\gamma]\cos\theta_1$
2	$(\mathrm{ch}\gamma - \cos\theta + \xi\,\mathrm{sh}\gamma)\theta/\sin\theta$	$[\mathrm{ch}(s-1)\gamma + \xi\,\mathrm{sh}(s-1)\gamma]\cos\theta_1$
3	$-(\mathrm{ch}\gamma^0 + \cos\theta + \xi\,\mathrm{sh}\gamma^0)\theta/\sin\theta$	$(-1)^{s-1}[\mathrm{ch}(s-1)\gamma^0 + \xi\,\mathrm{sh}(s-1)\gamma^0] \times\cos\theta_1$

Case	$N_s^{(r)}$ $(p = p_r)$
1, 2, 3	$(K_{s+1} - K_s\cos\theta_{s+1})/\sin\theta_{s+1}$, where $\theta_{s+1} = 0$ at $s = 1, ..., k-1$

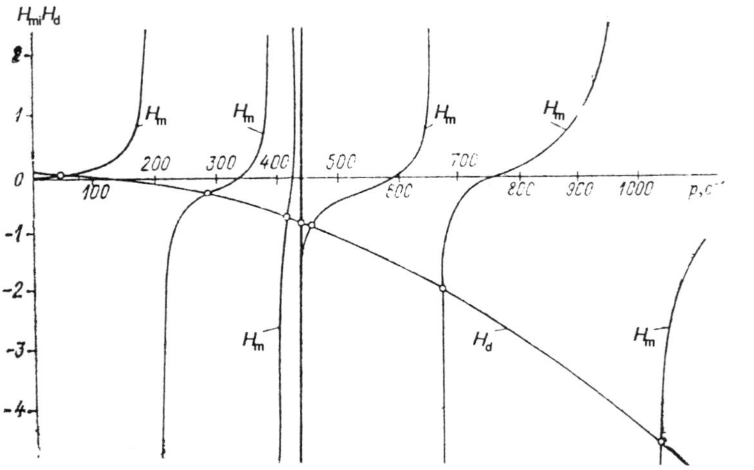

Figure 13. Curves $H_m(p)$ and $H_d(p)$.

1. For the fixed value p in the prescribed frequency range $\theta = \theta_s = p\Delta l/g$ and the functions $R_1 = A_1/B_1 = c_{13}(\Pi'_{12})^2 - J_{11}p^2$; $R_s = -p^2 J_{s3} \times (\Pi'_{s1})^2/(1 - p^2 J_{s3} c_{s2}^{-1})$ $(s \neq 1)$ are determined. The procedure of determining the functions A, B, C, D for the mechanisms of the analogous structure is illustrated in §7.

2. Determination of the function κ by (68).

3. Calculation of the function $H_m(p)$ by the formulae of Table 2 in accordance with the value κ.

4. Calculations of functions $H_d(p) = (R_1 - p\sigma^{-1}\mathrm{tg}\theta_1)/\Delta c_0$.

5. Determination of the "natural" frequencies as intersection points of the curves H_m (p) and H_d (p) or directly on a computer.

6. Calculation of stroboscopic modes of vibration K_s and loads N_s by formulae of Table 2 for the natural frequencies p_r.

7. Determination of the modal coefficients of the mode a_{sj} by formula (48) for an arbitrary element sj.

Plots of the functions H_m (p) and H_d (p) at $k = 6$ and $\Pi_{s1} = 0.5$ are presented in Fig. 13. In the root zone of the equation $R^{-1} = 0$, with the root equal to $p_* = \sqrt{c_{s2}/J_{s3}} = 457.9\ \mathrm{s}^{-1}$, density of distribution of the "natural" frequencies is substantially increasing.

The stroboscopic modes at $r = 1, ..., 7$ are presented in Fig. 14. The mode $r = 7$ $|H_d(p_*)| > |H_m(p_*)|$ corresponds to $r = 7'$ (see case 3), and at $|H_d(p_*)| < |H_m(p_*)|$ to $r = 7''$ (see case 2).

In this example the parameters of the system were slightly varied for detecting both varieties of the modes at $|\kappa| > 1$. It is of interest that at the "natural" frequencies beyond the point of thickening, the sequence of the stroboscopic modes, starting from the single node one, is repeated. The number of thickening points is equal to the number of roots of the equation $R_s^{-1}(p) = 0$.

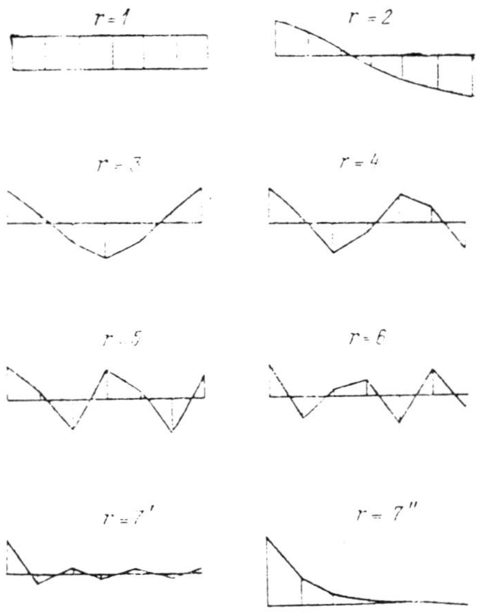

Figure 14. Stroboscopic forms of vibrations.

The influence of the number of mechanisms k on the value of the "natural" frequencies at the fixed value of the total moment of inertia of the driven members kJ_{s3} = const, and the overall stiffness coefficient of the driving mechanisms kc_{s2} = const, are shown in Table 3 (the case $k \to \infty$ will be analyzed in §12).

A particular case. At small values θ, the main shaft can be modelled as a subsystem with discrete parameters, with moment of inertia of the shaft segment ΔJ_0 being added to the moment of inertia of the corresponding driving member of the mechanism, and the torsional stiffness being considered by the coefficient $\Delta c_0 = GI/\Delta l$. The analysis shows that the inequality $kp^2 \Delta J_0 / \Delta c_0 \leq 0.1$–$0.2$ may serve as the criterion of smallness.

In this case, the relationships presented above are simplified, since it can be assumed that $\sin\theta \approx \theta$, $\mathrm{tg}\theta \approx \theta$, $\cos\theta \approx 1$. The resulting relationships are given

Table 3. Dependence of "natural" frequencies p_r on the number of identical mechanisms k.

k	r						
	1	2	3	4	5	6	7
2	48.10	278.8	386.4	—	—	673.8	1033
4	48.20	288.0	412.1	434.9	—	673.1	1032
6	48.30	291.0	416.3	440.3	459.2	672.5	1028
12	48.35	292.6	419.1	442.5	462.3	665.4	1020
∞	48.50	294.0	420.5	443.4	488.5	654.5	1012

Table 4. Relationships for determining frequency and modal characteristics of regular branched systems with concentrated parameters.

Case	$\varkappa$, γ, γ^0	$H_m(p)$	$u_s^{(r)}$ $(p=p_r)$	$Q_s^{(r)}$ $(p=p_r)$
1	$\vert\varkappa\vert<1$ $\gamma=\arccos\varkappa$	$\frac{\cos(k-0.5)\gamma}{\cos(k+0.5)\gamma}-1$	$\cos(s-1)\gamma+\mathrm{tg}(k+0.5)\gamma\sin(s-1)\gamma$ $(s=1, \ldots, n)$	$\Delta c_0\{\cos\gamma-\cos(s-1)\gamma+\mathrm{tg}(k+0.5\gamma)$ $\times[\sin s\gamma-\sin(s-1)\gamma]\}$
2	$\varkappa>1$ $\gamma=\mathrm{Arch}\varkappa$	$\frac{\mathrm{ch}(k-0.5)\gamma}{\mathrm{ch}(k+0.5)\gamma}-1$	$\mathrm{ch}(s-1)\gamma-\mathrm{th}(k+0.5)\gamma\,\mathrm{sh}(s-1)\gamma$ $(s=1, \ldots, n)$	$\Delta c_0\{\mathrm{ch}s\gamma-\mathrm{ch}(s-1)\gamma-\mathrm{th}(k+0.5)\gamma$ $\times[\mathrm{sh}s\gamma-\mathrm{sh}(s-1)]\}$
3	$\varkappa<-1$ $\gamma^0=\mathrm{Arch}\vert\varkappa\vert$	$-\left[\frac{\mathrm{sh}(k-0.5)\gamma^0}{\mathrm{sh}(k+0.5)\gamma^0}+1\right]$	$(-1)^{s-1}[\mathrm{ch}(s-1)\gamma^0-\mathrm{cth}(k$ $+0.5)\gamma^0\mathrm{sh}(s-1)\gamma^0]$ $(s=1, \ldots, n)$	$(-1)^s\Delta c_0\{\mathrm{ch}s\gamma^0+\mathrm{ch}(s-1)\gamma^0-\mathrm{cth}(k$ $+0.5)\gamma^0[\mathrm{sh}s\gamma^0+\mathrm{sh}(s-1)\gamma^0]\}$

in Table 4. It should be noted, however, that use of discrete representation of the main shaft usually does not lead to substantial computational savings.

Besides this method of describing this system, an algebraic presentation of the frequency equation, which is based on raising to power of transition matrices of the identical blocks, also can be used [15]. Then the formal frequency equation becomes

$$\sum_{\nu=0}^{k} \Delta c_0^{-\nu} R^{\nu-k}(\delta_{k\nu} R + \zeta_{k\nu} R_1) = 0 \tag{69}$$

where
$$\delta_{k\nu} = C_{k+\nu}^{2\nu+1} = (k+\nu)\ (k+\nu-1)\ \dots\ (k-\nu)/(2\nu+1)!$$

$$\zeta_{k\nu} = C_{k+\nu}^{\nu} = (k+\nu)\ (k+\nu-1)\ \dots\ (k-\nu+1)/(2\nu)$$

(C_n^m are combinations; $\nu = 0, 1, \dots, k$).

At the absolute rigid shaft $\Delta c_0 \to \infty$, considering that $\delta_{k0} = k$ and $\zeta_{k0} = 1$, it is easy to show that the formal frequency equation (69) is reduced to the form $R^{-(k-1)}(p)\ [k + R^{-1}(p)\, R_1(p)\ ] = 0$. This equation breaks into two equations $R^{-(k-1)} = 0$ and $k + R^{-1}R_1 = 0$, which in the adopted designations coincides with Eqs. (49) and (50) in §7.

§11. SIMPLIFIED METHOD FOR ANALYSIS OF FORCED VIBRATIONS

As it has been shown in §§4 and 7, in order to analyze the forced vibrations it is generally advisable to transform into quasinormal coordinates. However, in many practical applications the parameters of the drive change only slightly and, consequently, the "natural" frequencies $p(\tau)$ also change insignificantly. In such cases, on condition that the system is dynamically stable in all the frequency range (see §6), the analytical procedure can be substantially simplified. It is achieved by elimination of expansion of excitations by the modes of vibrations, with the calculations being performed using averaged values of parameters.

General case. Consider now the dynamic model in Fig. 9. First, consider the nonresonance conditions, for whose analysis the dissipative forces may be omitted without impairing the accuracy of the analysis. Suppose that the harmonic excitation force or torque $Q_i \cos\omega t$ is applied to the end of the branch i. Then on the arbitrary segment of the camshaft ($0 \le x_s \le \Delta l_s$) the torsional vibrations would be described by the following relationships: $\Delta\varphi_s = \tilde{a}_s(x_s)\cos\omega t$, where $\tilde{a}_s(x_s) = K_{s-1} \times \cos\vartheta_s + N_{s-1}\sin\vartheta_s$; $\vartheta_s = \omega x_s/g$; $g = \sqrt{GI/\rho}$; G is the shear modulus; I, ρ are the polar moment of the inertia of the cross section and the moment of the inertia of the unit of the shaft length. Repeating the calculations analogous to those in §7 for determination of the "natural" frequencies, we obtain for the harmonic process with the frequency ω the following recurrent relationships:

$$\left.\begin{aligned} K_s &= K_{s-1}\cos\theta_s + N_{s-1}\sin\theta_s \\ N_s &= \beta_s[-K_{s-1}\sin\theta_s + N_{s-1}\cos\theta_s + K_s R_s\sigma_s/\omega - \delta_s^i\sigma_s Q_i/(D_s\omega)] \end{aligned}\right\} \quad (70)$$

Here, $\theta_s = \vartheta_s(\Delta l_s)$; ϑ_s, β_s, σ_s, R_s, D_s (see §7); in all computational relationships p should be subsituted by ω; $\delta_s^i = 1$ at $s = i$; $\delta_s^i = 0$ at $s \neq i$ (δ_s^i is Kronecker's symbol of the second kind).

Relationships (70) differ from the structurally similar relationships (43) used before at the frequency analysis. Besides the formal substitution of p by ω, the difference is in the fact that at $s = i$ in the expression for N_s, the member appeared which is proportional to the amplitude of the exciting force Q_i; in addition, the amplitude of vibrations K_0 cannot be assigned arbitrarily.

Let us use the procedure of the "numerical experiment" [12], which is a modification of the method of the initial parameters [4]. The function N_n represents a linear combination relative to K_0 and Q_i

$$N_n = S_0(\omega)K_0 + S_i(\omega)Q_i \quad (71)$$

Let us perform two dummy calculations at $N_0 = 0$ by the algorithm presented in §7 and subject to the corrections following from the recurrent relationships (70). In the first caculation, assume $K_0^{(1)} = 1$, $Q_0^{(1)} = 0$. Then, from (71) $N_n^{(1)} = S_0$. In the second calculation, assume $K_0^{(2)} = 0$; $Q_i^{(2)} = 1$; hence $N_n^{(2)} = S_i$. At the chosen ω, after passing through all the segments, we obtain $N_n^{(1)}$ and $N_n^{(2)}$ as the known numbers. Since at the end of the camshaft the amplitude torque is equal to zero, we have $N_n = 0$. Hence from formula (71)

$$K_0 = -Q_i S_i/S_0 = -Q_i N_n^{(2)}/N_n^{(1)} \quad (72)$$

The ratio $-S_i/S_0$ may be treated as the harmonic coefficient of influence for the initial cross section $s = 0$. At $\omega \to p$ we have $S_0 \to 0$ and, consequently, $K_0 \to \infty$, which corresponds to the resonance if dissipation is neglected. Having determined from (72) K_0, and in accordance with the boundary conditions taking $N_0 = 0$, we find the amplitudes K_s by means of the recurrent relationships (70). In order to determine the amplitudes in the branches, the formula $\tilde{a}_{sj}^{(\omega)} = K_s\,\alpha_{sj}^{(\omega)}$ should be used, where $\alpha_{sj}^{(\omega)}$ is the coefficient determined from (48) after P_r is replaced with ω.

To estimate the resonance amplitudes, taking into consideration the integral character of information on the dissipative properties of the actual systems, it is advisable to use energy relationships. At the resonance of the mode r the exciting force $Q_i\cos\omega t$ during one period performs work $\Delta E_+^{(r)} = \pi Q_i \tilde{a}_{ij}^{(r)}$, where $\tilde{a}_{ij}^{(r)}$ is the amplitude of the forced vibrations of the element j of the branch $s = i$, to which the exciting force is applied.

Since at the resonance the amplitudes ratio is equal to the appropriate modal coefficient, we have $\tilde{a}_{ij}^{(r)} = a_{ij}^{(r)} K_0$, where $a_{ij}^{(r)}$ is the modal coefficient at $K_0 = 1$ (see §7). If the force is applied to the end of the branch, then $a_{ij}^{(r)} = K_i^{(r)}/D_i(p_r)$. The energy removed per one period due to dissipative forces is equal to $\Delta E_{-}^{(r)} = 0.5\psi_r K_0^2 J_r p_r^2$, where J_r is determined by formula (53) at $K_r^* = \alpha_r^* = 1$.

The condition of balance between the supplied and removed energy $\Delta E_{+}^{(r)} = E_{-}^{(r)}$ considering the presented relationships, results in the following relationship: $K_0 = 2\pi a_{ij}^{(r)} Q_i / (\psi_r J_r p_r^2)$. For calculation of the amplitude in the arbitrary cross section, K_0 should be multiplied by the corresponding modal coefficient $a_{sj}^{(r)}$. The obtained result is easily generalized for the case when exciting forces are applied in many points of the system. This corresponds to the substitution $a_{ij}^{(r)} Q_i$ by the sum $\Sigma a_{ij}^{(r)} Q_{ij}$ in all cross sections for which $Q_{ij} \neq 0$.

The described procedure can be expanded to the general case of the periodic exciting force by using Fourier series and method of superposition.

Model of regular structure. In §10 the frequency analysis of a regular system of branched mechanisms has been considered. Now, the forced vibrations of the analogous system will be studied, and first of all the non-resonance case is considered. Suppose the harmonic inducing force (torque) $M = Q \cos\omega t$ (see Fig. 12) is applied to the ends of the branches $s = 2, ..., n$. The system (70) will be valid if the last term of the second equation is equal to $\sigma_s, Q_s:(D_s\omega)$. In our case, due to regularity of the system, $A_s, B_s, C_s, D_s, \sigma_s, \theta_s, \beta_s = 1$ do not depend on s. Therefore, at these parameters the subscript s may be omitted. Relationships (70) (subject to the mentioned corrections) represent a heterogeneous system of differential equations, whose solution is sought in the form $K_s = K_s^* + K_s^{**}$; $N_s = N_s^* + N_s^{**}$. Here, the solution of the homogeneous system of equations is indicated by one asterisk, and particular solution of the heterogeneous system of equations is indicated by two asterisks. Taking the particular solution of the homogeneous equation as $K_s^* = C_k^0 \lambda^s$, $N_s^* = C_N^0 \lambda^s$, where C_k^0, C_N^0 are the arbitrary constants, we obtain the characteristic equation whose solution is

$$\lambda = \varkappa \mp \sqrt{\varkappa^2 - 1}$$

where $\varkappa = \cos\theta - 0.5 Z \sin\theta$; $Z = -\sigma C/(D\omega)$.

Using the obtained values of λ it is possible to show that in the general format the functions K_s and N_s are described as

$$\left.\begin{aligned} K_s &= h_1 f_s + h_2 v_s + K^{**} \\ N_s &= \sin^{-1}\theta [h_1(f_{s+1} - f_s\cos\theta) + h_2(v_{s+1} - v_s\cos\theta)] + N^{**} \end{aligned}\right\} \quad (73)$$

Here, $h_1(\omega)$, $h_2(\omega)$ are some functions determinable by boundary conditions.

Depending on the value of the parameter $\varkappa$, which is determined from formula (68) at $p = \omega$, five cases can be chosen; for each of them the functions f_s, v_s, K_s^{**}, N_s^{**} are presented in Table 5.

The absence of the amplitude loads at the ends, which is provided at $N_0 = 0$ and $N_n = 0$, serves as boundary conditions. Subject to (70), we obtain a heterogeneous system of linear algebraic equations with two unknowns h_1 and h_2:

$$\left.\begin{aligned} h_1(f_{n+1} - f_n\cos\theta) + h_2(v_{n+1} - v_n\cos\theta) &= -K^{**}(1 - \cos\theta) \\ h_1(f_2 - \cos\theta - U\theta^{-1}\sin\theta) + h_2 v_2 &= K^{**}[U\theta^{-1}\sin\theta - (1 - \cos\theta)] \end{aligned}\right\} \quad (74)$$

Table 5. Relationships for calculating forced vibrations of regular systems

<table>
<tr><th>Case</th><th>$\varkappa$</th><th>f_s</th><th>v_s</th><th>K^{**}</th><th>N^{**}</th></tr>
<tr><td>1</td><td>$|\varkappa|<1$
$(\varkappa=\cos\gamma)$</td><td>$\cos(s-1)\gamma$</td><td>$\sin(s-1)\gamma$</td><td rowspan="3">$\dfrac{\zeta Q}{2\mathrm{tg}\,\dfrac{\theta}{2}+Z}$</td><td rowspan="5">$K_s^{**}\,\mathrm{tg}\,\dfrac{\theta}{2}$</td></tr>
<tr><td>2</td><td>$\varkappa>1$
$(\varkappa=\mathrm{ch}\,\gamma)$</td><td>$\mathrm{ch}(s-1)\gamma$</td><td>$\mathrm{sh}(s-1)\gamma$</td></tr>
<tr><td>3</td><td>$\varkappa<-1$
$(|\varkappa|=\mathrm{ch}\,\gamma)$</td><td>$(-1)^{s-1}$
$\times\mathrm{ch}(s-1)\gamma$</td><td>$(-1)^{s-1}$
$\times\mathrm{sh}(s-1)\gamma$</td></tr>
<tr><td>4</td><td>$\varkappa=1$</td><td>1</td><td>$s-1$</td><td>$-0.5\zeta Q s^2\sin\theta$</td></tr>
<tr><td>5</td><td>$\varkappa=-1$</td><td>$(-1)^{s-1}$</td><td>$(-1)^{s-1}(s-1)$</td><td>$-0.25\zeta Q\sin\theta$</td></tr>
<tr><td colspan="6">Note: $\zeta=\sigma/(D\omega)$; $s>1$.</td></tr>
</table>

Here, $U = R_1/\Delta c_0$; $R_1 = A_1/B_1$ at the fixed end of the branch $s = 1$, and $R_1 = C_1/D_1$ at the free end; $\Delta c_0 = GI/\Delta l$ is the torsional stiffness of the segment of the shaft.

Having substituted the roots of Eqs. (74) into relationship (73) for K_s, we find the amplitude-frequency characteristic on the distribution shaft. The amplitudes of forced vibrations in the branches are determined in the same way as for a general case.

The resonance conditions for a regular system are calculated by the method presented above. However, it should be taken into consideration that the exciting force is applied to the ends of all regular branches, which leads to the following formula:

$$\tilde{a}_{ij}^{(r)} = 2a_{ij}^{(r)}\pi Q \sum_{s=2}^{n} a_{sj}^{(r)} f(\varphi_r, J_r, p_r^2)$$

§12. CONTINUAL MODEL OF DRIVE AT UNIFORM DISTRIBUTION OF DYNAMIC CHARACTERISTICS OF MECHANISMS ALONG AXIS OF THE MAIN SHAFT

Preliminary notes. The complexity of modern machines and great dynamic coupling of their individual units and mechanisms cause a need for consideration of vibratory systems of large dimensions. In analysis and, especially, in dynamic synthesis of such systems, major difficulties arise when the set of generalized coordinates and variable parameters becomes difficult to survey. In order to overcome these difficulties, in [10, 17, 21] *continual models* were proposed as applied to the problems of dynamics of mechanisms, in which the kinematic, elastic, and inertial properties of mechanisms are represented by some integral characteristics. It allows for use of a generalized representation of the group of variables, substantially reducing the number of coordinates characterizing the vibratory system and simplifying its analysis and synthesis. Now the analogous procedure (aggregation) is used wider and wider for studies of complex objects of mechanics, automatic control, and economics [4].

As a carrier of dynamic properties of mechanisms, some pseudomedium of a vibratory structure can be chosen. Depending on the characteristic properties of the system, the media of various structures are proposed. They differ from a simple medium considered in classical mechanics since a number of degrees of freedom more than three is assigned to each point. The review of such media and the analysis of one model of a medium of complex structure are presented in the monograph [38].

The difference between the pseudomedium used for the description of the complex systems of mechanisms and the known media are, first of all, due to consideration of kinematic characteristics which are predetermining nonstationary properties of this medium.

Investigation methods of global drive model by means of continual idealization procedure. Let's concretize now the procedure of application of the pseudomedium in the problems of mechanism dynamics. As an example, consider the regular system of the mechanisms branching off the main shaft (see §10). As before, the main shaft *1* (Fig. 15) is presented in the form of a torsional subsystem with distributed parameters, to whose right end the driving mechanism *3* ($s = n$)[1] is attached: from the main shaft k, identical mechanisms are branching off which are presented as a chain of discrete elements *2*. Now, let s represent k identical mechanisms as a pseudomedium. This medium is formed by the "spreading" of elastic, inertial, and kinematic characteristics along the shaft axis and has the property of transferring force and motion only along the direction corresponding to the transmission of the motion by the mechanism. Interaction of "vertical" chains of the elements of the medium takes place only via the main shaft subsystem.

[1] In the model considered in §10, the driving mechanism was attached to the left end ($s = 0$). This circumstance, of course, cannot influence the results of the analysis since only the direction of counting is changing. By a lucky choice of the direction of counting it is often possible to obtain more laconic calculation relationships. This is why the place of attachment of the driving mechanism would vary in the further chapters in an appropriate way.

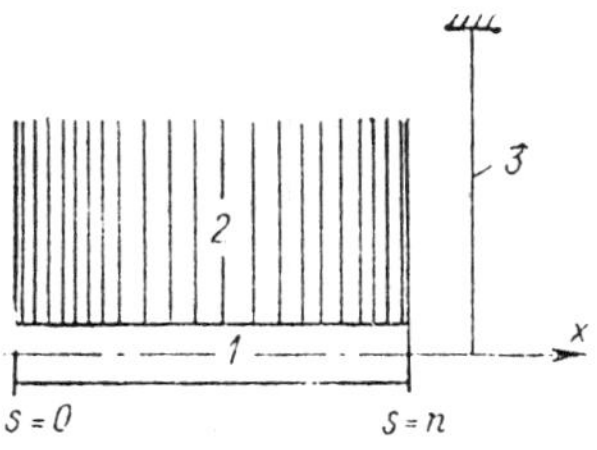

Figure 15. Continual dynamic model of a drive of a branched structure.

For a regular system of identical mechanisms, the "density" of the medium is constant, which corresponds to the uniform distribution along the the axis x. A more complex case of nonuniform distribution will be considered in §13.

The distributed modified transition matrix of the mechanism

$$\tilde{\Gamma} = \prod_{j=j_{max}}^{1} \tilde{\Gamma}_j$$

serves as the characteristic of the medium. The matrix, as before, is formed as the product of the transition matrices of the elements $\tilde{\Gamma}_j$ (see §7) in a reverse order. The only difference is that when determining $\tilde{\Gamma}_j$ for the inertial and elastic elements, these elements should be previously distributed along the axis x. In this case, when distribution is uniform, the corresponding characteristics must be referred to the unit of length. For example

$$\tilde{\Gamma}_j^{(c)} = \begin{bmatrix} 1 & l/(c_j k) \\ 0 & 1 \end{bmatrix} \qquad \tilde{\Gamma}_j^{(J)} = \begin{bmatrix} 1 & 0 \\ -p^2 J_j k/l & 1 \end{bmatrix}$$

Here, c_j is the stiffness coefficient; J_j is the moment of inertia; p is the "natural" frequency; l is the length of the main shaft.

The following partial differential equation (with dissipative members being omitted) serves as a mathematical model on the basis of which the frequency and modal properties of the considered system can be determined,

$$\rho_0 \partial^2 \varphi / \partial t^2 - G I_0 \partial^2 \varphi / \partial x^2 - S(\varphi, \tau) = 0 \tag{75}$$

Here, $\varphi(x, \tau)$ is the additional twisting of the cross section x caused by the elastic deformations of the system elements; I_0, ρ_0 is the polar moment of the inertia of the cross section and the mass moment of the inertia of the unit of length; G is the shear modulus; S is the reactive torque acting from the mechanisms on the unit of length.

Represent now the particular approximate solution of Eq. (75) as $\varphi = X(x, \tau)\,\Psi(t)$ where τ is "slow" time. Accordingly, fast and slow components of the function $S = \mu(\tau)\, X(x, \tau)\, \Psi(t)$ are to be separated; here, $\mu = -C/D$, where C, D are the elements of the second line of the matrix $\tilde{\Gamma}$. After the substitution of φ and S into Eq. (75), considering the slowness of the change of the amplitude values X and μ, we obtain

$$X'' + P(\tau) X = 0 \tag{76}$$

$$\ddot{\Psi} + \frac{\dot{\mu}}{\mu}\dot{\Psi} + p^2(\tau)\,\Psi = 0 \tag{77}$$

where $P = (\rho_0 p^2 + \mu)/(G I_0)$; $(\ )' = \partial/\partial x$; $(\dot{\ }) = \partial/\partial t$.

In Eq. (76) the function μ is the characteristic of the medium corresponding to the mechanisms. It enters into the coefficient P, which can be given the form $P = \rho p^2/(G I_0)$, where $\rho = \rho_0 + \mu p^{-2}$. The considered mathematical model corresponds to some fictitious shaft possessing the fictitious moment of inertia, $\tilde{J} = \rho l$, which can be positive, negative, zero, or infinity depending on the value p. On the other hand, if the component related to μ were separated from the coefficient P, then in Eq. (76) the member $\mu X/(GI_0)$ would appear, which corresponds to some fictitious elastic moment. The coefficient $\mu/(GI_0)$ plays the role of the distributed dynamic stiffness. Further, we will use both the first and the second analogy.

Let's introduce now the following functions:

$$f=\begin{cases}\cos\theta(x,\tau) & \text{at} \quad P>0\\ \operatorname{ch}\theta(x,\tau) & \text{at} \quad P\leqslant 0\end{cases} \qquad v=\begin{cases}\sin\theta(x,\tau) & \text{at} \quad P>0\\ \operatorname{sh}\theta(x,\tau) & \text{at} \quad P\leqslant 0\end{cases} \tag{78}$$

where $\theta(x,\tau)=\tilde{\lambda}(\tau)x$; $\tilde{\lambda}(\tau)=\sqrt{|P|}$.

Then the solution of Eq. (76) can be written as:

$$X=h(\tau)[f(\theta)+\alpha(\tau)v(\theta)] \tag{79}$$

$$X'=h(\tau)\tilde{\lambda}[\partial f/\partial\theta+\alpha(\tau)\partial v/\partial\theta] \tag{80}$$

with $\partial v/\partial\theta = f$, and $\partial f/\partial\theta = -v$ at $P > 0$, and $\partial f/\partial\theta = v$ at $P \leq 0$.

Substitute the boundary condition $X'(0) = 0$ into Eq. (80). Since $f(0) = 1$; $v(0) = 0$, we have $\alpha \equiv 0$. Since the mode of vibrations is determined up to the constant multiplier, assume $X(l) = 1$; hence $h(\tau) = f^{-1}(l, \tau)$. In addition to the conditions mentioned, we have one more condition, that is the equality of the amplitude torques in the cross section $x = l$ both for the considered subsystem and for the driving mechanism. This condition serves as a formal frequency equation

$$\Phi(p) = H_m(p) - H_d(p) = 0 \tag{81}$$

where $H_m(p) = lX(\theta_*)/X(\theta_*) = \theta_* \partial f/\partial\theta(\theta_*)/f(\theta_*)$; $H_d(p) = -lR_n/(GI_0) = -c_0^{-1}R_n$. Here c_0 is the torsional stiffness of the shaft; $\theta_* = \theta(l, \tau)$.

It is easy to prove that $H_m = -\theta_{*r}\operatorname{tg}\theta_{*r}$ at $P > 0$, and $H_m = \theta_{*r}\operatorname{th}\theta_{*r}$ at $P \leq 0$, where r is the number of the mode; $\theta_{*r} = \theta_*$ at $p = p_r$. The modes of vibrations are determined by relationship (79) at the substitution of the "natural" frequencies $p = p_r$, which are the roots of Eq. (81). Here $X_r = \cos\theta_r/\cos\theta_{r*}$ at $P(p_r) > 0$, and $X_r = \operatorname{ch}\theta_r/\operatorname{ch}\theta_{*r}$ at $P(p_r) \leq 0$.

The determination of nonstationary forms in the branches and the transition to the local model are carried out in the same manner as before (see §7) at $K_{0s} = X_0(x_s)$.

The analysis of the formal frequency equation (81) allows to reveal some characteristic domains of parameter changes.

Domain 1. $P > 0$ ($\rho > 0$). Here, the mode of vibrations is described by trigonometric functions.

Domain 2. $P < 0$ ($\rho < 0$), $H_d > 0$. The mode of vibrations is described by hyperbolic functions, with the amplitude decreasing with increasing distance from the place of attachment of the drive (spatial damping).

Domain 3. $P < 0$ ($\rho < 0$), $H_d < 0$. In the corresponding frequency range Eq. (81) does not have any roots since $\operatorname{sgn}H_d = -\operatorname{sgn}H_m$.

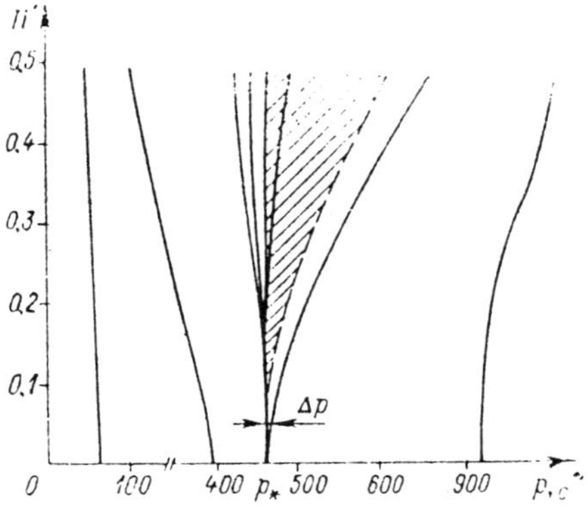

Figure 16. Curves $\Pi'(p_r)$.

Fig. 16, for the example considered in §10, shows curves $\Pi'(p_r)$, illustrating evolution of the changes of the spectrum of the "natural" frequencies depending on the values of the first kinematic transfer function Π' at $\Pi' \in [0, 0.5]$. In the Figure, domain 2 is shown by hatching; in this range $P \leq 0$ and the solution is described by hyperbolic functions. The other frequency ranges correspond to domain 1. An especially intensive concentration of the frequency spectrum takes place in the zone Δp in the vicinity of the point of thickening $p_* = 457.9s^{-1}$; here, the infinite number of "natural" frequencies is located. It should be remembered that the point of thickening corresponds to the partial frequency of the subsystem of mechanisms when their input members are fixed. The infinite number of frequencies in this zone, of course, is due to continual idealization; when the number of mechanisms k is limited, the number of the mode achievable in this zone is equal to $k - 1$. The hyperbolic shape of the modes of vibrations is detected immediately beyond the zone of thickening. The analogous "spatial" damping also was revealed by V. Palmov when investigating vibratory media in which an infinite set of noninteracting isotropic oscillators with continously distributed natural frequencies [38] is connected with each point of the carrying medium (in our case the main shaft acts as a carrying medium). After the hyperbolic mode of vibrations there is a repetition of modes on the main shaft beginning with the second (single-node) mode. However, it does not mean the full identity of the dynamic picture observable in the system before the point of thickening, since more complex modes appear in this frequency range in the pseudomedium itself, the latter corresponding to the vibratory contour of mechanisms.

Connection between the continual model and regular structure models. It is obvious that the continual model considered above is a limiting case for the regular system of mechanisms (see §10) at the number of mechanisms k $\to \infty$. Let's establish now the connection between the characteristic index λ of the system of differential equations (61) and the parameter $\tilde{\lambda}$. Single out in the regular system one block s consisting of one mechanism s which is characterized by "reaction" $R_s = C_s / D_s$, and a segment Δl_s of the shaft. As it has been established in §10, $\lambda_s = \cos\gamma_s + i\sin\gamma_s\,(i=\sqrt{-1})$, with $\cos\gamma_s = \cos\theta_s + 0.5\,\sigma p^{-1}R_s\sin\theta_s$ (here θ_s corresponds to the notation adopted in §10). At small γ_s and θ_s we have $\cos\gamma_s \approx 1 - 0.5\gamma_s^2$, $\cos\theta_s \approx 1 - 0.5\theta_s^2$. Consequently, $\gamma_s \approx \theta_s\sqrt{1 - R_s\Delta J_s^{-1}p^{-2}}$. On the other hand, it is easy to prove that the same result corresponds to $\tilde{\lambda}\Delta l_s$. Thus, $\gamma_s \approx \tilde{\lambda}\Delta l_s$; this means that the mode of vibrations in the continual model coincides with the stroboscopic mode of vibrations at sufficiently small segments Δl_s between the input members of the mechanisms. Of course, this conclusion could be made from purely physical premises.

For the example considered above in §10 (see Table 3), the comparison of "natural" frequencies for a number of modes was performed at the different number of the identical mechanisms k and the same value of the total moment of inertia and the total reduced stiffness of all the mechanisms. While analyzing the table, it is easy to see a good numerical correlation of the results regardless of the number of mechanisms k. This confirms the closeness between the stroboscopic modes and the mode obtained on the continual model $k \to \infty$, and also shows the effectiveness of this model. In the zone near the point of thickening, the disagreements increase a little. There are also losses of a qualitative character. For example, the number of modes in the zone depends on the number k, whereas at the continual idealization it is infinity. In addition, the given model excludes the possibility of detecting the modes of vibrations described by hyperbolic functions with the phase shift π at the transition from one mechanism to another (case 3, $\varkappa < -1$) since at the continual idealization $\Delta l_s \to 0$.

Generalized continual model of a mixed type. Let's look now at the dynamic model of the branched structure drive presented in Fig. 17. The model consists of n segments, each of them corresponding to the pseudomedium with its own dynamic characteristics. The subsystems of mechanisms whose dynamic characteristics are specified in discrete format are located on the boundary of two segments (in the figure these mechanisms are shown by bold lines). A need for analysis of such systems arises in cases of series connection of sharply differing subsystems (e.g., stepped shaft), as well as in cases when the system, together with a great number of homogeneous mechanisms, has some "powerful" mechanisms whose influence should be separated. A role of such mechanisms could be played by driving mechanisms located in the intermediate cross sections. This allows for consideration of the issue of the rational (from the dynamics standpoint) positioning of the attachment place of the drive to the camshaft.

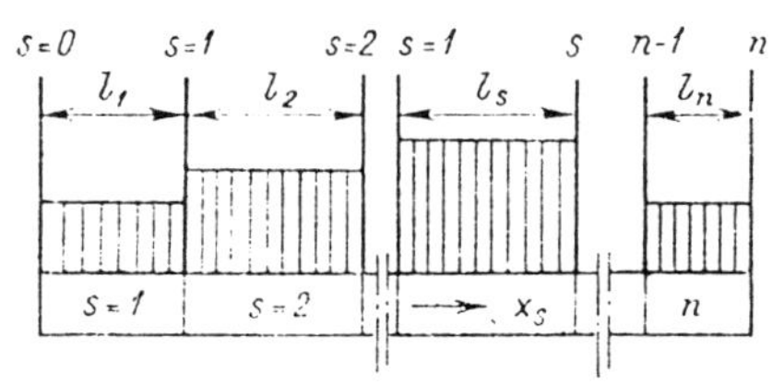

Figure 17. Generalized model of discrete-continual type.

For a segment s located between the cross sections $s-1$ and s it can be written, based on formulas (79) and (80), that

$$X_s = h_s(f_s + \alpha_s v_s); \; X'_s = h_s \theta_s l_s^{-1}(\partial f_s/\partial \theta_s + \alpha_s \partial v_s/\partial \theta_s) \qquad (82)$$

Here, f_s and υ_s are determined by formulae (78) at the argument $\theta_s = \tilde{\lambda}_s x_s$ depending on the value of the function $P_s = \rho_s p^2/(GI_s)$ ($\rho_s = \rho_{0s} + \mu_s p^{-2}$); the reference point for x_s is the initial cross section of the segment s ($0 \le x_s \le l_s$).

The conditions of matching at the boundaries of the segments can be represented as

$$h_s = h_{s-1}(f_{s*} + \alpha_s v_{s*}) \qquad (83)$$

$$H^-_{s+1} c_{s+1} = R_s + H^+_s c_s \qquad (84)$$

Here, $H^+_s = l_s X'_s(l_s)/X_s(l_s) = \theta_{s*}[(\partial f/\partial \theta_s)_* + \alpha_s(\partial v_s/\partial \theta_s)_*]/(f_{s*} + \alpha_s v_{s*})$ $H^-_{s+1} = l_{s+1} X'_{s+1}(0)/X_{s+1}(0) = (\theta_{s+1})_* \alpha_{s+1}$; $R_s = A_s/B_s$ at the fixed end of the branch, and $R_s = C_s/D_s$ at the free end of the branch; $c_s = GI_s/l_s$ is the torsional stiffness coefficient of the segment s; the functions corresponding to $x_s = l_s$ are designated by an asterisk; the superscript "minus" at the function H corresponds to the left boundary of the segment s, and the superscript "plus" corresponds to its right boundary.

The "natural" frequencies p_r are determined on the basis of (83) and (84) using the boundary conditions. Here, one value h_s may be chosen at will, and two boundary conditions allow finding out one value α_s and the unknown frequency p_r. As an example, consider a model of the drive schematized as two segments on whose boundaries the driving mechanism is located. In the segment $s = 0$ the amplitude torque is zero, therefore $H^-_1 = \theta_{1*}\alpha_1 = 0$. Since $\theta_{1*} = \tilde{\lambda}_1 l_1 \neq 0$, we have $\alpha_1 = 0$. Then, according to (83) and taking $h_1 = 1$, we obtain $1 = h_0 f_{1*}$ or $h_0 = f^{-1}_{1*}$. On the boundary of the segments ($s = 1$) relationship (84) is valid, $H^-_2 c_2 = R_1 + H^+_1 c_1$ at $H^-_2 = \theta_{2*}\alpha_2$, $H^+_1 = \theta_{1*} f^{-1}_{1*} (\partial f_1/\partial \theta_1)_*$ ($\alpha_1 \equiv 0$). Also, at the right end of the segment $s = 2$ $H^+_2 = 0$; this condition is satisfied at $\alpha_2 = -f^{-1}_{2*} (df_2/d\theta_2)_*$. After substitution of the expressions for H^-_2 and H^+_1 into Eq. (84) at $s = 1$, we obtain the formal frequency equation as

$$R_1 \pm c_1 \theta_{1*} v_{1*} f^{-1}_{1*} \pm c_2 \theta_{2*} v_{2*} f^{-1}_{2*} = 0 \qquad (85)$$

Here, the "plus" sign corresponds to $P_s \le 0$, and the "minus" sign corresponds to $P_s > 0$.

It is easy to make sure that at the absence of one of the segments ($l_1 = 0$ or $l_2 = 0$) Eq. (85) coincides with Eq. (81) obtained before. The mode of vibrations for the example considered above is described as follows:

$$X_1 = f_1/f_{1*}, \qquad X_2 = f_2 \pm v_{2*} v_2/f_{2*}$$

It should be noted, in conclusion, that the presented procedure is very simple and allows, with primitive computational means (e.g., a hand-held calculator) to carry out quickly not only the analysis of complex systems, but formation of frequency spectra as well.

§13. CONTINUAL MODEL OF DRIVE WITH NONUNIFORM DISTRIBUTION OF DYNAMIC CHARACTERISTICS OF MECHANISMS ALONG AXIS OF MAIN SHAFT

In the preceding section we considered the case of the uniform distribution of mechanism characteristics along the axis x. Now we are going to consider a more complex case in which the pseudomedium corresponding to the dynamic characteristics of mechanisms has a changing density of distribution along the axis x [10]. The latter may occur either with the identical mechanisms distributed nonuniformly along the main shaft or at different mechanisms whose "contribution" towards frequency spectrum may vary. In the latter case, the density of distribution depends also on the frequency range analyzed.

The initial premises of the analysis are presented in §12. The difference in this case is the dependence on x of the coefficient at X in Eq. (76):

$$X'' + P(x;\ \tau)X = 0 \tag{86}$$

where $P = [\rho_0 p^2 + \mu(x,\ \tau)]/(GI_0)$.

If we assume $\tilde{\rho} = \rho_0 + \mu/p^2$ to be a fictitious distributed characteristic of the moment of inertia along the axis x, then the corresponding fictitious moment of inertia will be $\tilde{J} =$

$$\int_0^x \tilde{\rho}\,dx \quad \text{at} \quad \tilde{\rho} \approx \rho$$

A sufficiently simple and effective method of distribution is the choice of $\tilde{\rho}\,(x)$ in a certain class of the smooth functions. The mean value of the curve $\rho\,(x)$ and the invariability of the location of the "center of gravity" of this curve have to be competed with. Then

$$\left.\begin{aligned} \bar{\rho}l &= \int_0^l \tilde{\rho}(x,\ \tau;\ \zeta_1,\ \ldots,\ \zeta_N)dx = J_0 - p^{-2}\sum_{s=1}^{n} R_s \\ \rho l x_c &= \int_0^l x\tilde{\rho}(x,\ \tau,\ \zeta_1,\ \ldots,\ \zeta_N)dx = 0.5J_0 l - p^{-2}\sum_{s=1}^{n} x_s R_s \end{aligned}\right\} \tag{87}$$

where l is the length of the shaft; $\bar{\rho}$ is the mean value of ρ; x_c is the coordinate of the "center of gravity" of $\rho\,(x)$; J_0 is the moment of inertia of the main shaft; $R_s = C_s/D_s$; $\zeta_1, \ldots, \zeta_N$ are the free parameters of the family of functions.

If $N > 2$, then $N - 2$ parameters can be used to satisfy the additional conditions of distribution or be chosen arbitrarily.

The application of the condition $\lambda\Delta l_s \approx \gamma_s$, obtained in §12, gives a little more accurate result. In order to determine γ_s, the main shaft is divided into n segments (according to the number of mechanisms). For each segment we find $\cos\gamma_s = \varkappa_s$ at $|\varkappa_s| < 1$ or $\mathrm{ch}\gamma_s = \varkappa_s$ at $\varkappa_s > 1$, where $\varkappa_s = \cos\theta_s + 0.5\,\sigma p^{-1} R_s \sin\theta_s$ [see (68)]. Then $P(x)$ is determined as a stepped piece-meal constant function at $P_s = \pm\gamma_s^2 p^{-2}\Delta l_s^{-2}$, with the "plus" sign corresponding to $|\varkappa_s| < 1$, and the "minus" sign corresponding to $\varkappa_s > 1$. This function can be replaced by a smooth function analogous to $\rho\,(x)$ by means of the method shown above.

In order to solve the differential equation (86) we use one family of the exact

Table 6. Calculation relationships for type characteristics ρ (x).

Function	"Slow" change of ρ (x)	Family 1	Family 2
ρ	$\rho(x)$	$\zeta_1\bar{\rho}(\zeta_2 x/l+1)^{-4}$	$\zeta_1\bar{\rho}(\zeta_2 x/l+1)^{-2}$
ξ	$\sqrt[4]{\rho_0/\rho}$	$\zeta_2 x/l+1$	$\sqrt{\zeta_2 x/l+1}$
ρ_0	$\rho(0)$	$\frac{(1+\zeta_2)^3}{1+\zeta_2+\zeta_2^2/3}\bar{\rho}$	$(1+\zeta_2)\bar{\rho}$
ϑ	$px\sqrt{\rho(0)/(GI_0)}$	$px\sqrt{\zeta_1\bar{\rho}/(GI_0)}$	$px\sqrt{\zeta_1\bar{\rho}/(GI_0)}$
θ	$\frac{p}{\sqrt{GI_0}}\int_0^x\sqrt{\rho}\,dx$	$\vartheta\left(\zeta_2\frac{x}{l}+1\right)^{-1}$	$\frac{\vartheta\sqrt{1-0.25\zeta_2^2\vartheta^{-2}}}{\zeta_2}\times\ln\left(\zeta_2\frac{x}{l}+1\right)$
θ_*	$\theta(l)$	$\theta(l)$	$\theta(l)$
$\tilde{l}_c=\frac{x_c}{l}$	$\frac{1}{l^2\bar{\rho}}\int_0^l x\rho\,dx$	$\frac{3+\zeta_2}{6(1+\zeta_2+\zeta_2^2/3)}$	$\frac{1+\zeta_2}{\zeta_2^2}\left[\ln(1+\zeta_2)-\frac{\zeta_2}{1+\zeta_2}\right]$

solutions of the equation for the conditional oscillator. Then the function X for the arbitrary mode r may be presented as:

$$X=\xi(x)[h_1(\tau)\cos\theta(x)+h_2(\tau)\sin\theta(x)]$$

where h_1 (τ), h_2 (τ) is determined by boundary conditions,

$$\xi\ (x)=\exp\ [-0.5\ (z-z_0)];\ \theta(x)=\sqrt{lP(0)l}\int_0^x \exp z dx$$

z is the particular solution of the equation of the conditional oscillator.

In Table 6 a number of characteristics are specified for the case of the "slow" change of the function ρ (x) and for two families of the functions, each of which depends on two parameters ζ_1 and ζ_2.

In order to explain the calculation procedure, we will use the family of functions 1 when the function $P(x) = p^2\rho(x)$ can be approximated as $P(x) = p^2 (\zeta_2 x/l + 1)^{-4}\zeta_1\bar{\rho}$. First we find the parameter ζ_2 on the basis of the last formula of Table 6, using $\tilde{l}_c = x_c/l$ from (87).

$$\zeta_2=0.25[\tilde{l}_c^{-1}-6+\sqrt{12(\tilde{l}_c^{-1}-1)+\tilde{l}_c^{-2}}]$$

At $\tilde{l}_c = 0.5$ we have $\zeta_2 = 0$, which corresponds in the first approximation to the uniform distribution (see §12). Term $\rho(0)/\bar{\rho}$ acts as the parameter ζ_1; its value in Table 6 is expressed in terms of ζ_2. In Fig. 18 the curves are shown which simplify selection of these parameters and illustrate the nature of their influence on ρ (x).

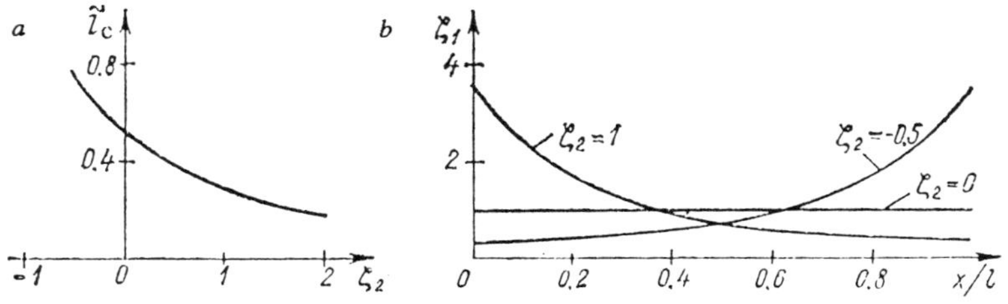

Figure 18. Characteristics of distribution of function $\rho(x)$ of continual model.

Suppose that the driving mechanism is located in the cross section $x = 0$. Then the formal frequency equation is reduced to the format of (81) at $H_d(p) = c_0^{-1} R_0$, where c_0 is the torsional stiffness of the shaft; $R_0 = A_0/B_0$ at the fixed input cross section of the driving mechanism, and $R_0 = C_0/D_0$ at the free input cross section (see §7). The relationships for $H_m(p)$ and the modes of vibrations X_r obtained considering the formulae of Table 6 and the boundary conditions are presented below.

Domain 1. $\bar{\rho} > 0$:

$$H_m = \{\mathrm{tg}\theta_*[\theta_*(1+\varsigma_2)+\varsigma_2^2\theta_*^{-1}]-\varsigma_2^2\}/(1+\varsigma_2\theta_*^{-1}\mathrm{tg}\theta_*)$$

$$X_r = (\varsigma_2 x/l+1)[\cos\theta+\sin\theta(\theta_*\mathrm{tg}\theta_*-\varsigma_2)/(\theta_*+\varsigma_2\mathrm{tg}\theta_*)]_{p=p_r}$$

In Fig. 19*a* plots $H_m(\theta_*, \zeta_2)$ are presented. At $\zeta_2 = -\theta_*: \mathrm{tg}\theta_*$ we have $H_m \to \infty$. The roots of this equation correspond to the transition boundary (by θ_*) from one mode of vibrations to another. Since $\theta_* = pl\,(1 + \zeta_2)^{-1}\sqrt{\zeta_1\bar{\rho}/(GI_0)}$ (see Table 6) it turns out that at $\rho \to \infty$, respectively, $\theta_* \to \infty$. In the vicinity of the corresponding value p_* the transition condition from one mode to another is satisfied many times and, consequently, a narrow frequency range appears in which the infinite number of "natural" frequencies and modes are assembled. The corresponding frequency p_* was qualified above as a point of thickening. It is possible to show that at $r \gg \pi^{-1}\sqrt{(H_m + \zeta_2^2)/(1 + \zeta_2)}$ (r is the modal number) the influence of the driving mechanism on the "natural" frequencies becomes negligible.

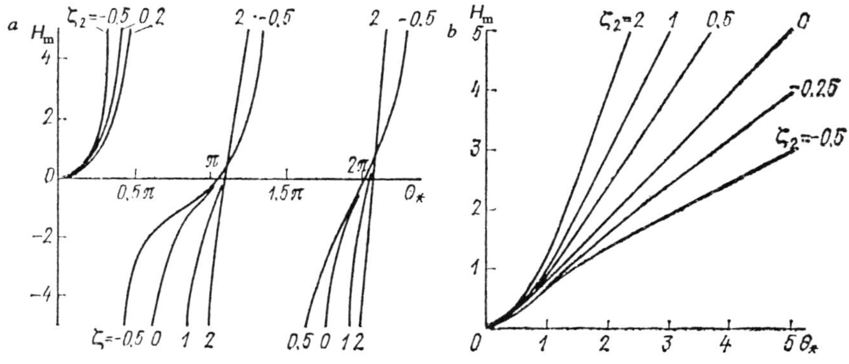

Figure 19. Plots $H_m(\theta_*, \zeta_2)$.

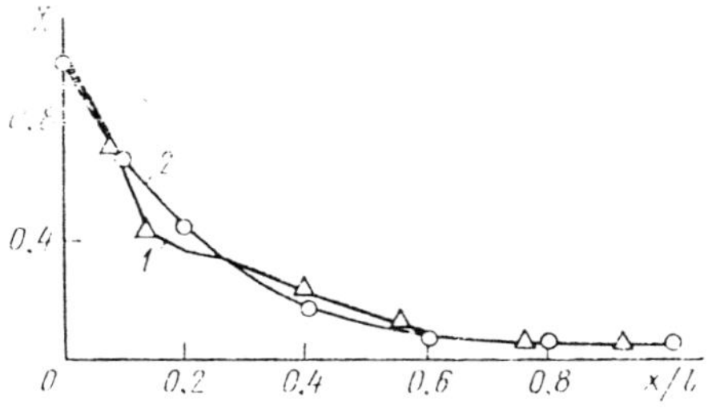

Figure 20. The exponential stroboscopic form of vibrations for continual and discrete models.

Domain 2. $\rho < 0$.

$$H_{\mathrm{m}} = \{\mathrm{th}\,\theta_* [\zeta_2 \theta_*^{-1} - \theta_* (1 + \zeta_2)] - \zeta_2\} (1 + \zeta_2 \theta_*^{-1} \mathrm{th}\,\theta_*)$$

$$X_r = (\zeta_2 x/l + 1)[\mathrm{ch}\,\theta - \mathrm{sh}\,\theta (\theta_* \mathrm{th}\,\theta_* + \zeta_2)\,(\theta_* + \zeta_2 \mathrm{th}\,\theta_*)]|_{\rho = \rho_r}$$

The plots of the functions $H_{\mathrm{m}}(\theta_*)$ at different values of parameters ζ_2 are presented in Fig. 19*b*. Fig. 20 shows the vibration modes which are characteristic for the considered domain shown.

Curve 1 corresponds to the case of 12 discrete specified mechanisms whose parameters correspond to the example considered in §7 in the vicinity of the first point of thickening; curve 2 is for continual representation.

The curve presented illustrates well the effect of “spatial damping”.

CHAPTER
THREE

RING STRUCTURE DRIVES

§14. DYNAMIC MODEL OF A RING STRUCTURE DRIVE WITH DISCRETE PARAMETERS

Dynamic model of the first level ("global model"). In §1 the examples were presented when, in order to move massive operating organs, a number of identical mechanisms operating in parallel are used. In such cases, statically indeterminable systems of the ring structure develop. In this section we are going to consider the dynamic model with three actuators and an elastic drive in which all the inertial elements are specified as discrete ones (Fig. 21). The similar drives of the operating members are used in looms, knitting machines, etc. It should be noted that a number of drives most commonly used in paractice can be analyzed as particular cases of the considered model. For example, $c_5 = 0$ corresponds to the system with six degrees of freedom at two actuators; at $c_7 = c_8 = 0$ the number of degrees of freedom is also equal to six, but the model corresponds to three branched mechanisms; at $c_5 = 0$, $J_2 = 0$, $J_5 = 0$ we have a model with four degrees of freedom; at $c_1 = c_2 = c_3 = \infty$ the model corresponds to the absolutely rigid drive and the distributing shaft (camshaft).

The global model (see §4) corresponding to the considered drive will have the form presented in Fig. 21 if we ignore the dissipative elements and external disturbances. When analyzing this model we are going to use the system of the linearized equations (8). In the drives represented by models of large dimensions, the apparatus of modified transition matrices will be used (see below).

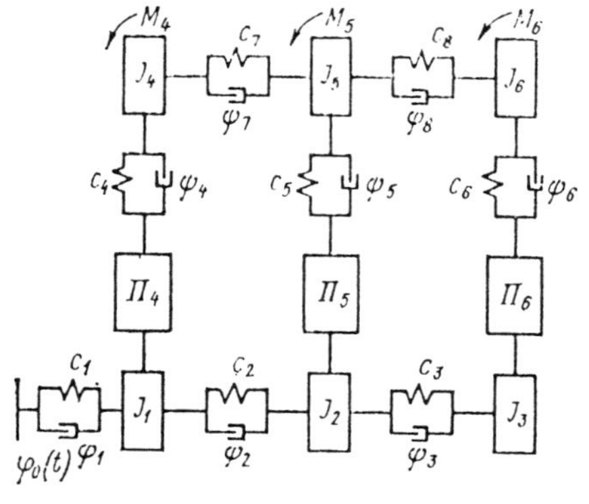

Figure 21. Dynamic model of ring structure with discrete elements.

The considered vibratory system has six degrees of freedom. Having determined the inertial and quasielastic coefficients $a_{ij}^*(t)$ and $c_{ij}^*(t)$, the formal frequency equation is written in the form (22) as follows

$$\begin{vmatrix} c_1+c_2+c_4\Pi_*'^2-J_1p^2 & -c_2 & 0 & -c_4\Pi_*' & 0 & 0 \\ -c_2 & c_2+c_3+c_5\Pi_*'^2-J_2p^2 & -c_3 & 0 & -c_5\Pi_*' & 0 \\ 0 & -c_3 & c_3+c_6\Pi_*'^2-J_3p^2 & 0 & 0 & -c_6\Pi_*' \\ -c_4\Pi_*' & 0 & 0 & c_4+c_7-J_4p^2 & -c_7 & 0 \\ 0 & -c_5\Pi_*' & 0 & -c_7 & c_5+c_7+c_8-J_5p^2 & -c_8 \\ 0 & 0 & -c_6\Pi_*' & 0 & -c_8 & c_6+c_8-J_6p^2 \end{vmatrix}=0 \quad (88)$$

After solving this equation we find six values of the "natural" frequencies p_r. The nonstationary modal coefficients α_{ir} can be found from the system of algebraic equations (25). In this case, for each fixed value of the "natural" frequency p_r, the system of five equations of the form (25) should be written, from which we find five values α_{ir} [during derivation of equations (25) it was assumed that $\alpha_{1r} = 1$]. Since $r = 1, \ldots, 6$, the number of such systems is equal to six and the full number of the sought modal coefficients at the fixed value Π_*' is equal to 30. It is obvious that similar calculations should be performed on computers.

Dynamic model of the second level ("local" model). In this case we have six "local" models described in quasinormal coordinates η_r by heterogeneous differential equations of a second order with varying coefficients (27). The moment of inertia, reduced to the mode r, is equal to

$$J_r = \sum_{i=1}^{6} \alpha_{ir}^2 J_i$$

reduced stiffness coefficient $c_r = p_r^2 J_r$; the reduced dissipative coefficient $\overline{b}_r = \psi_r p_r J_r/(2\pi)$, where ψ_r is the reduced coefficient of dissipation (§4); from (29), $\tilde{b}_r = \omega_0 \partial J_r/\partial\varphi_0$. On the basis of (5) and (27), the reduced exciting forces are described by relationships

$$F_r = \sum_{i=1}^{6} Q_i^* \gamma_{ir}$$

and $\dot{\varphi}_0 = \omega_0 = \text{const}$; for $i = 1, 2, 3$, it follows that $Q_i^* = 0$, $i = 4, 5, 6$, $Q_i^* = -M_i - J_i \times \omega_0^2 \Pi_*''$.

After solving the differential equation (27), the reverse transformation to the original generalized coordinates is performed by the relationships

$$q_i = \sum_{r=1}^{H} \gamma_{ir} \eta_r \quad (i = 1, \ldots, H)$$

Conditions of quasi-stationarity. In §6 it has been shown that even at the slow change of parameters it is possible to lose dynamic stability in the last time interval, which is associated with an undesirable increase of vibroactivity of the drive. In the ring structure system the source of the parameter variability is the function Π_*', the same for all actuators. Accordingly, the condition (35) can be reduced to

$$\psi_r > 2\pi\omega_0 p_r^{-1} \beta_{r\ \max} |\Pi_*''|_{\max} \tag{89}$$

where ψ_r is the value of the dissipation coefficient reduced to the mode r (see §4); $\omega_0 = \dot{\varphi}_0 = \text{const}$ is the angular velocity of the ideal drive; $\beta_r = p_r^{-1} \partial p_r/\partial\Pi_*'$ is the coefficient of nonstationarity characterizing the intensity of the change of the "natural" frequencies with changing Π_*'.

If the possibilities of reducing the coefficient of nonstationarity by means of the rational choice of the system parameters are exhausted, then in accordance with (89) the maximum angular velocity of the drive can be established at which the dynamic stability in any segment of the kinematic cycle is assured.

Distribution of kinetostatic loads between actuators in an ideal drive. Issues related to the problem of load distribution in mechanisms forming statically indeterminate systems were partly considered in [11, 26, 27, 28]. For the original model (see Fig. 21), the distribution of kinetostatic loads at the absence of clearances and errors of manufacturing and assembly is described by the set of linear algebraic equations

$$\sum_{i=1}^{6} c_{ji} q_i = Q_j^* \qquad (j = 1, \ldots, 6) \tag{90}$$

Here, $Q_j^* = 0$ at $j = 1, 2, 3$, and $Q_j^* = -M_j - J_j \Pi_*'' \omega_0^2$ at $j = 4, 5, 6$. The coefficients c_{ij} are equal to the elements of the row j and the column i of the determinant (88) at $p \equiv 0$.

Such an approach, of course, is valid only for those harmonics of functions Q_j^* which are substantially smaller than the lowest "natural" frequency. This corresponds to the kinetostatic analysis. Even with ideal manufacturing of mechanisms, the load can be distributed extremely unevenly. In order to obtain the desirable distribution of the loads, we may use the system of equations (90) and solve it to find the required stiffness ratios.

Consideration of clearances and manufacturing errors. For a rational design of statically indeterminate kinematic chains, it is usually necessary to take account not only of elastic deformations, but also of the manufacturing errors and kinematic backlashes (clearances). The problem can be solved by considering three different position functions, Π_4, Π_5, Π_6 (the subscript corresponds to the subscript of the adjoining elastic element; see Fig. 21). In addition, it is assumed that each actuator has clearances. Then the system of differential equations, ignoring the dissipative forces, becomes

$$\left.\begin{aligned}
&J_1\ddot{q}_1+(c_1+c_2)q_1-c_2q_2-c_4\Pi'_{4*}\Delta\varphi_4=0\\
&J_2\ddot{q}_2-c_2q_1+(c_2+c_3)q_2-c_3q_3-c_5\Pi'_{5*}\Delta\varphi_5=0\\
&J_3\ddot{q}_3-c_3q_2+c_3q_3-c_6\Pi'_{6*}\Delta\varphi_6=0\\
&J_4\ddot{q}_4+c_4\Delta\varphi_4-c_7\Delta\varphi_7=-M_4-J_4\Pi''_{4*}\omega_0^2\\
&J_5\ddot{q}_5+c_5\Delta\varphi_5+c_7\Delta\varphi_7-c_8\Delta\varphi_8=-M_5-J_5\Pi''_{5*}\omega_0^2\\
&J_6\ddot{q}_6+c_6\Delta\varphi_6+c_8\Delta\varphi_8=-M_6-J_6\Pi''_{6*}\omega_0^2
\end{aligned}\right\}\qquad(91)$$

and at $i = 4; 5; 6$

$$\Delta\varphi_i=\begin{cases} q_i-\Pi'_{i*}q_{i-3}-\Delta_i, & \text{if}\quad q_i-\Pi'_{i*}q_{i-3}>\Delta_i\\ q_i-\Pi'_{i*}q_{i-3}+\Delta_i, & \text{if}\quad q_i-\Pi'_{i*}q_{i-3}<\Delta_i\\ 0, & \text{if}\quad |q_i-\Pi'_{i*}q_{i-3}|\leqslant\Delta_i\end{cases}$$

at $i = 7; 8$

$$\Delta\varphi_7=q_5-q_4+\Pi_{5*}-\Pi_{4*}$$

$$\Delta\varphi_8=q_6-q_5+\Pi_{6*}-\Pi_{5*}$$

Here, $2\Delta_i$ is the angular displacement of the driven system corresponding to the clearance in the mechanism i.

The system can be solved by numerical methods, but at certain values of the criterion K_Δ the influence of clearances would be reduced, in general, to excitations of the pulse character (see §6).

If in the system (91) we assume $\ddot{q}_j \equiv 0$, then the obtained system of algebraic equations gives the possibility of finding the distribution of the kinetostatic loads, considering clearances and errors of realization of geometric characteristics. Conditions of continuity of the kinematic chain can also be determined. So, for example, for two mechanisms operating in parallel and at $Q_4^* = Q_5^* = Q^*$. $\Delta_i = \Delta$, and $c_1 \to \infty$, this condition, ignoring manufacturing inaccuracies, takes the form $|\, Q^* \,| > c_7 \Delta / (1 + 2c_7 / c_2)$. In order to reveal the role of the mechanism i in the considered closed contour, assume that $c_i = 0$ in the system (90) and determine for this case the value $\Delta\varphi_i^0 = [q_i - \Pi'_{i*}\, q_{j-3}]\; c_i = 0$. Further, comparing $\Delta\varphi_i^0$ and Δ_i, we can establish the relative duration of time intervals $\varepsilon = \Delta t/\tau$, (where $\tau = 2\pi/\omega_0$) on which $|\, \Delta\varphi_i^0 \,| < |\, \Delta_i \,|$. The value ε would characterize the section of the kinematic cycle in which the given mechanism does not take the kinetostatic load. By the rational choice of stiffnesses it is possible to get such a design that on the greater part of the cycle the drive would operate without switching of the working surfaces in the kinematic pairs. In this sense, the systems with a slight preload should be considered prospective. The preload can be judiciously provided at the design stage, for example, by introducing a small regulated phase shift of the functions Π_i corresponding to different mechanisms.

§15. GENERALIZED DYNAMIC MODEL WITH RING STRUCTURE MECHANISMS

Dynamic model of the first level ("global" model). Consider now a sufficiently general schematic (Fig. 22*a*), when motion from drive *1* is transferred to main (distributing) shaft *2*, from which a number of mechanisms such as gear trains, linkages, cams, etc. are branching off. Besides branches of type *3* considered in Chapter 2, there are mechanisms of type *5* which, together with the operating members *4* and the main shaft 2, form statically indeterminate vibratory systems of the ring structure.

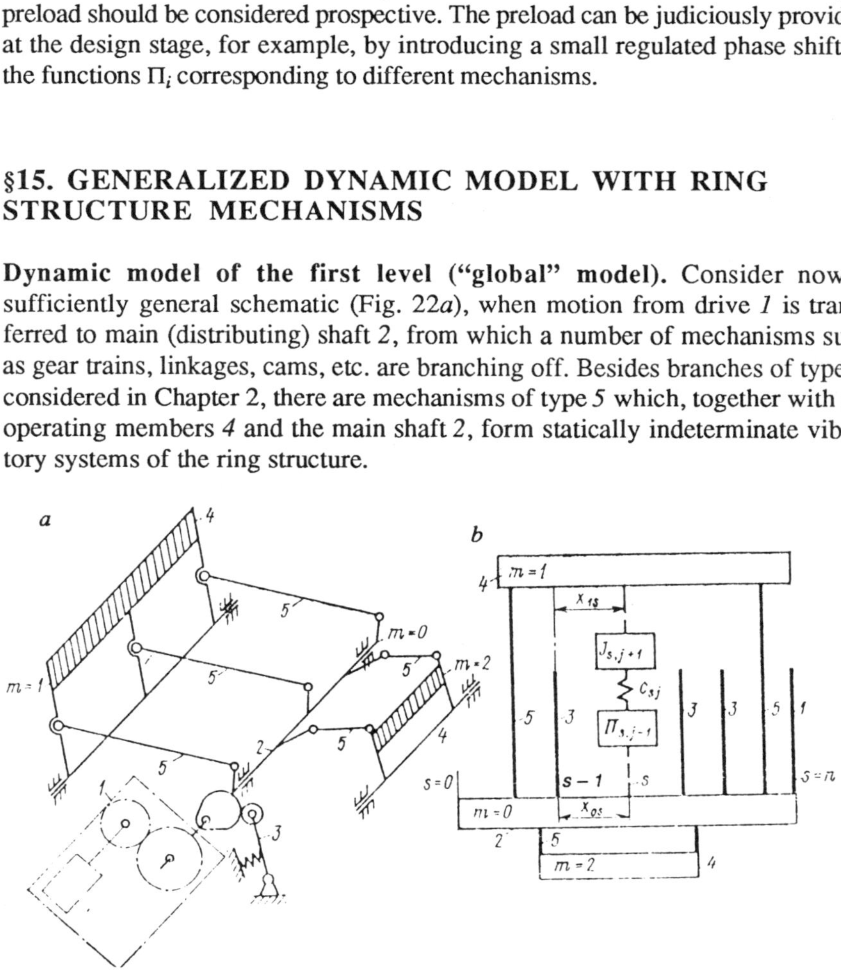

Figure 22. The machine drive and its dynamic model.

As a "global" model, the system in Fig. 22*b* is assumed to consist of a number of torsional vibratory subsystems $m = 0, 1, \ldots, m_*$ with distributed parameters corresponding to the main shaft *2* ($m = 0$) and to the operating members, and it also consists of a number of subsystems with lumped parameters which correspond to driving mechanism *1*, simple branches *3*, and mechanisms *5* operating in a parallel circuit and entering into the closed contours.

As in Chapter 2, the direction of counting of the elements is assumed from the main shaft, the subscript s indicates the number of the chain when passing along the main shaft from left to right, and the subscript j indicates the ordinal number of the element in the given chain with the adopted counting system. As an example, the elements $(s, j-1)$, (sj), and $(s, j+1)$ are shown, which can be kinematic, elastic, and inertial. The segment s of the subsystem with distributed parameters is located between the cross sections of the input members of the branches $s-1$ and s. In the block s (see §8), in addition to the mechanism s, the segments s of each of the subsystems adjoining to it are also included.

For mathematical description of the model we will use the approach given in detail in §§7 and 8. As in the drives of the branched structure, the nonstationary mode of vibrations in each segment s of the subsystem with the distributed parameters m is described as:

$$X_{ms}(x_{ms}, \tau) = K_{m,s-1}\cos\theta_{ms} + N_{m,s-1}\sin\theta_{ms} \tag{92}$$

where $\theta_{ms} = p(\tau)x_{ms}/g_{ms}$; $g_{ms} = \sqrt{GI_{ms}/\rho_{ms}}$; $0 \leq x_{ms} \leq \Delta l_{ms}$; G is shear modulus; I_{ms}, ρ_{ms} is the polar moment of inertia and moment of inertia of the mass of the unit length in the segment s of the subsystem m; p is the "natural" frequency; τ is "slow" time.

The amplitude value of the load for each mode is proportional to $Q_{ms} = GI_{ms}\, \partial X_{ms}/\partial x_{ms}$. It is easy to find out that $K_{ms} = X_{ms}(\Delta l_{ms}, \tau)$ makes up the stroboscopic mode of vibrations, and N_{ms} is proportional to the stroboscopic mode of the loads.

The following recurrent relationships are valid for the coefficients K_{ms}, N_{ms}:

$$\left.\begin{aligned} K_{ms} &= K_{m,s-1}\cos\theta_{ms} + N_{m,s-1}\sin\theta_{ms} \\ N_{ms} &= \beta_{ms}(-K_{m,s-1}\sin\theta_{ms} + N_{m,s-1}\cos\theta_{ms} + Z_{ms}) \end{aligned}\right\} \tag{93}$$

where $\beta_{ms} = \sigma_{m,s+1}/\sigma_{ms}$ $(\beta_{mn} = 1)$; $\sigma_{ms} = (GI_{ms}\rho_{ms})^{-0.5}$; $\theta_{ms} = \theta_{ms}(\Delta l_{ms})$.

The functions Z_{ms}, which are proportional to the amplitudes of the reactive moments applied in segment s to subsystems m, are determined depending on the form of the attachment of the second end of the branch s using formulae in Table 7. Each case is identified by the number υ_{ms}.

If the chain s has intermediate branches, they may be taken into account by means of the structural transformations presented in §8 by introducing the corresponding fictitious moment of inertia in the place of the attachment of this branch to the chain s.

Table 7. Functions Z_{ms}.

ν_{ms}	Z_{0s} $(m=0)$	Z_{ms} $(m=1,\ldots,h)$	Connection of chain
0	0	0	No connection of the chain with the subsystem m
1	$\dfrac{\sigma_{0s}C_sK_{0s}}{pD_s}$	$\dfrac{\sigma_{ms}C_sK_{ms}}{pA_s}$	The end of chain s is free
2	$\dfrac{\sigma_{0s}A_sK_{0s}}{pB_s}$	$\dfrac{\sigma_{ms}D_sK_{ms}}{pB_s}$	The end of the chain s is fixed
3	$\dfrac{\sigma_{0s}(A_sK_{0s}-K_{us})}{pB_s}$	$\dfrac{\sigma_{us}}{pB_s}(D_sK_{us}-K_{0s})$	The connection of the subsystem $m=0$ with the subsystem $m=u$ by means of the mechanism s

Notes: A_s, B_s, C_s, D_s are elements of the transition matrix of the mechanism s.

Proceed now with considering the boundary conditions. Suppose that the driving mechanism is attached to the distributing shaft in the cross section n. The absence of loads on all the ends of the subsystems with distributed parameters corresponds to the following conditions: $N_{m0} = 0$ and $N_{mn} = 0$ at $m = 0, 1, \ldots, m_*$. However, in order to satisfy these conditions, on the basis of recurrent relationships (93) we must assign the stroboscopic mode in the initial cross section, i.e., the values K_{m0}. In comparison with the analogous situation in the drives of the branched structure (see §7), an additional difficulty arises due to the fact that only one value K_{m0} for example, $K_{00} = 1$ may be chosen independently; the other values K_{m0} ($m = 1, \ldots, m_*$), as well as the frequency p, must be determined from the boundary conditions given above. The analytical expressions for the function K_{m0} in such systems are quite cumbersome and not very suitable for calculation on a computer. In addition, they change their format depending on the structure of the system. Therefore we use a more convenient "numerical experiment" approach [12] which already has been used in Chapter 2. It is obvious that in the considered linear problem the functions N_{mn} are linear combinations of K_{m0}, therefore at $N_{m0} = 0$ we have the following set of equations:

$$S_{m0}(p)K_{00}+S_{m1}(p)K_{10}+\ldots+S_{mm_*}(p)K_{m_*0}=N_{mn} \tag{94}$$

$$(m = 0, 1, \ldots, m_*)$$

where $S_{m0}, S_{m1}, \ldots, S_{mm^*}$ are the unknown coefficients.

Let's assume now the following matrix of the fictitious boundary conditions: $K_{m0}^{(k)}=\varkappa_{mk}$

$$\underset{[(m_*+1)\times(m_*+1)]}{\|\varkappa_{mk}\|}=\begin{bmatrix}1 & 0 \ldots 0\\ 0 & 1 \ldots 0\\ \ldots & \ldots\\ 0 & 0 \ldots 1\end{bmatrix}$$

where $k = 0, 1, \ldots, m_*$ is the number of the count.

It is easy to see that at the fictitious boundary conditions each equation of the system (94) gives $S_{mk} = N^{(k)}_{mn}$. Thus, in order to determine all the unknown coefficients S_{mk}, the calculation by recurrent relationships (93) should be carried out $m_* + 1$ times. Each count has a corresponding column of boundary conditions in the matrix $\| \kappa_{mk} \|$. After this, one of the equations of the system (94), for example, the equation $m = 0$, should be isolated and the real boundary conditions $k_{00} = 1$ and $N_{mn} = 0$ substituted into the remaining system m_* of equations. By solving this set of equations, true values of $K_{10}, ..., K_{m^*0}$ (m_* values altogether) are found.

An example would illustrate determination of the unknown boundary conditions. Suppose that the number of the subsystems linked with the main shaft is two ($m_* = 2$). Then for a given p, three calculations should be performed at the fictitious boundary conditions, namely, at $K^{(0)}_{00} = 1, K^{(0)}_{10} = K^{(0)}_{20} = 0$ ($k = 0$); at $K^{(1)}_{00} = 0, K^{(1)}_{10} = 1, K^{(1)}_{20} = 0$ ($k = 1$); and at $K^{(2)}_{00} = K^{(2)}_{10} = 0, K^{(2)}_{20} = 1$ ($k = 2$). Each count gives three values $N^{(k)}_{mn} = S_{mk}$. The unknown real boundary conditions K_{10} and K_{20} are determined from the set of two equations:

$$\left.\begin{aligned} S_{11}K_{10} + S_{12}K_{20} &= -S_{10} \\ S_{21}K_{10} + S_{22}K_{20} &= -S_{20} \end{aligned}\right\}$$

With one subsystem of the ring structure ($m_* = 1$) $K_{10} = -S_{10}/S_{11}$, where $S_{10} = N^{(0)}_{1n}$ at $k = 0$ ($K^{(0)}_{00} = 1, K^{0}_{10} = 0$), and $S_{11} = N^{(1)}_{1n}$ at $k = 1$ ($K^{(1)}_{00} = 0, K^{(1)}_{10} = 1$).

In order to find the unknown frequency p we now have several options. For example, the equation $m = 0$ at N_{0n} isolated before may serve as a formal frequency equation, namely:

$$U(p) = \sum_{m=0}^{m_*} S_{0m}(p) K_{m0}(p) = 0 \tag{95}$$

Here, S_{0m} and K_{m0} for the given p have been determined before by the numerical experiment.

Eq. (95) is transcendental in essence and is solved by the numerical method, e.g., by the step-by-step passage of the given frequency range to determine values $p = p_r$ corresponding to the sign change of the function $U(p)$.

Another form of the frequency equation is more convenient for the dynamic synthesis of the system and forming the desired frequency characteristics. It can be obtained in the following manner: having determined $K_{m0}(p)$ from the system (94), we find the function $H_m(p) = K_{0,n-1} \sin\theta_{0n} - N_{0,n-1} \cos\theta_{0n}$ using relationship (93).

Since $N_{0n} = 0$, it is easy to see that the function $H_m = H_d$, where $H_d = Z_{0n}(p)$. Formulae for Z_{0n} are given in Table 7, with $\nu_{0n} = 1$ or $\nu_{0n} = 2$ depending on the form of the fixation of the second end of the driving mechanism.

Thus, the "natural" frequencies p_r are determined as the abscissae of the intersection points of two curves, $H_m(p)$ and $H_d(p)$.

The stroboscopic mode of vibrations is characterized by the values $K_{ms}(p_r)$, where p_r is the "natural" frequency. In the arbitrary cross section of the subsystems m, the nonstationary mode of vibrations is described by relationship (92).

For the arbitrary element *sj* (branch *s*, ordinal number *j*) the modal coefficients should be determined by the formulae:

$$a_{sj}^{(r)} = [A_{sj}^{*} K_{0s} - B_{sj}^{*} Z_{0s} p \sigma_{0s}^{-1}]_{p=p_r} \quad \textbf{at} \quad \nu_{0s} \neq 0 \tag{96}$$

at $\nu_{0s} = 0$, $\nu_{ms} \neq 0$

$$a_{sj}^{(r)} = \begin{cases} [A_{sj}^{*} C_s K_{ms} A_s^{-1}]_{p=p_r} & (\nu_{ms} = 1) \\ [B_{sj}^{*} K_{ms} B_s^{-1}]_{p=p_r} & (\nu_{ms} = 2) \end{cases} \tag{97}$$

where A_{sj}^{*}, B_{sj}^{*} are the elements of the first row of the matrix $\Gamma_{sj}^{*} = \prod_{u=j}^{1} \Gamma_{su}$.

Just as in analysis of branched systems, in order to avoid degeneracy of modes at some values of parameters, e.g., in dwelling periods of driven members of some mechanisms, we may use the normalized values of the nonstationary modal coefficients $\zeta_{sj}^{r} = a_{sj}^{(r)} / a_{*}^{(r)}$. Here, $a_{*}^{(r)}$ corresponds to some specified element or cross section, for selection of which the condition $|\zeta_{sj}^{r}| \leq 1$ is satisfied. Analysis shows that relatively great values $|\zeta_{sj}^{r}|$, representing strong coupling, correspond to the mechanisms with close partial frequencies.

In the real drives it often happens that $\theta_{ms} << 1$. It means that the subsystems could be represented as a set of discrete elastic and inertial elements. Here, $\cos\theta_{ms} \approx 1$; $\sin\theta_{ms} \approx \theta_{ms} = p\sqrt{\Delta J_{ms}/\Delta c_{ms}}$; $\sigma_{ms} = 1/\sqrt{\Delta J_{ms} \Delta c_{ms}}$; $g_{ms} = \Delta l_{ms} \sqrt{\Delta c_{ms}/\Delta J_{ms}}$, where ΔJ_{ms}, Δc_{ms} are the moment of inertia and the coefficient of torsional stiffness of the segment *s* of the subsystem *m*, respectively. It should be noted, however, that transition to the discrete parameters does not give appreciable simplification of computations.

Dynamic models of the second level ("local" models). The local model represents the system with a single degree of freedom with changing parameters, which is described by the differential equations (27) and (33) in the quasinormal coordinates. The moment of inertia J_r reduced to the mode *r*, and the right-hand side of the differential equation (33) in this case become:

$$\left.\begin{aligned} J_r = \Big\{ & 0.5(a_{*}^{(r)})^{-2} \sum_{s=1}^{n} \sum_{m=0}^{m_*} \rho_{ms} [(K_{m,\,s-1}^2 + N_{m,\,s-1}^2) \Delta l_{ms} \\ & + 0.5 g_{ms} p^{-1} (K_{m,\,s-1}^2 - N_{m,\,s-1}^2) \sin 2\theta_{ms} \\ & + g_{ms} p^{-1} K_{m,\,s-1} N_{m,\,s-1} (1 - \cos 2\theta_{ms})] \\ & + \sum_{s=1}^{n} \sum_{j=1}^{j_{\max}} J_{sj} (\varsigma_{sj}^{(r)})^2 \Big\}_{p=p_r} \\ W_r = J_r^{-1} & \sum_{s=1}^{n} \Big\{ \sum_{j=1}^{j_{\max}} (F_{sj} - J_{sj} \ddot{\varphi}_{sj}^{*}) \zeta_{sj}^{r} \\ & + (a_{*}^{(r)})^{-1} \sum_{m=0}^{m_*} \int_0^{\Delta l_{ms}} [\mu_{ms}(x_{ms}, t) \end{aligned}\right\} \tag{98}$$

$$- \rho_{ms} \ddot{\varphi}^*_m(t)] X_{ms}(x_{ms}, \tau) dx_{ms} \Big\} \Big|_{p=p_r} \qquad (98\ Cont.)$$

Here, F_{sj} is the torque or force applied to the element *sj*; φ^*_{sj}, φ^*_m are the absolute coordinates of the element *sj* and of the subsystem *m* in the programmed motion (i.e., ignoring elasticity of members); μ_{ms} is the external distributed torque applied to the subsystem *m*.

In many specific cases, the subintegral functions entering in the expression W_r are quite simple. For example, at $\theta_{ms} << 1$ the function X_{ms} is close to the linear one $X_{ms} \approx K_{m,\,s-1} + N_{m,\,s-1} \vartheta_{ms}$, where $\vartheta_{ms} = p_r x_{ms},\ g_{ms}$.

The dissipation coefficient ψ_r reduced to the mode *r* is determined in the same way as in § 7, with the only difference that the number of the subsystems with distributed parameters is equal to $m_* + 1$,

$$\psi_r = \frac{\sum_{s=1}^{n} \left(\sum_{m=0}^{m_*} \psi_{ms} V_{ms} + \sum_{j=1}^{j_{max}} \psi_{sj} V_{sj} \right)}{\sum_{s=1}^{n} \left(\sum_{m=0}^{m_*} V_{ms} + \sum_{j=1}^{j_{max}} V_{sj} \right)} \Bigg|_{p=p_r}$$

Here, ψ_{ms}, ψ_{sj} are the dissipation coefficients of the segment *s* of the subsystem *m* and of the element *sj*; V_{ms}, V_{sj} is potential energy corresponding to these elements (see §7). At $\psi_{ms} = \psi_{sj} = \psi^*$ we have $\psi_r = \psi^*$.

The refinement of these values ψ_r for nonsingle frequency vibrations, considering nonlinearity of dissipative characteristics, are presented in [11, 37].

§16. MODEL OF DRIVE FOR TRANSLATIONAL PROGRAMMED MOTION OF HEAVY WORKING MEMBERS

In a number of technological machines for performing translational programmed motion of massive driven members, identical mechanisms operating in parallel arrangement are used. Such arrangements are widely used for driving heavy tables, carriages, etc. in automatic book sewers, combing machines, rolling mills, and in other machines. The drive shown in Fig. 23 serves as an example of such an arrangement. In programmed motion (ignoring the elasticity of members) the driven member *6* performs a strictly translational motion. However, due to elasticity, this motion happens to be just planar.

For schematization of the drive, as in §15, consider the main shaft as a torsional subsystem with the distributed parameters, and all the mechanisms including the driving one as subsystems with the discrete (lumped) parameters. The massive driven member *6* is considered to be a rigid body of mass m_0 with moment of inertia relative to the center of mass *J*.

Let us use the results of the analysis from §15. Assume that $s = 1$ corresponds to the driving mechanism, and $s = 2, \ldots, n$ are actuators connected with the common driving member. In the case considered, the number of the driven subsystems forming closed contours is one ($m_* = 1$). Besides, the mode of vibra-

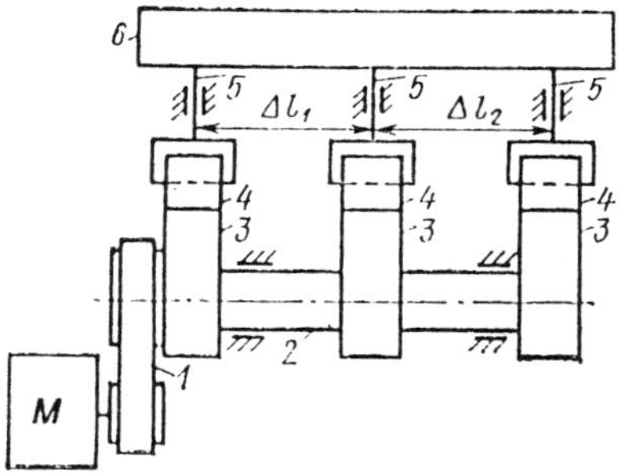

Figure 23. The drive for translational programmed motion of a heavy operating member M is the motor; *1*) belt transmission; *2*) camshaft; *3*) cam; *4*) roller; *5*) tapper; *6*) working member.

tions of the driven member $X_1 = Y(x)$ is represented by a linear relationship from the coordinate. Here,

$$K_{1s} = Y_c + \alpha \tilde{x}_s \tag{99}$$

where $\tilde{x}_s = \sum_{i=1}^{s} \Delta l_i - x_c$; x_c is the coordinate of the center of masses; $Y_c = Y(\tilde{x}_c)$; α is the amplitude value of the angular displacement of the driven member.

The stroboscopic mode of the main shaft K_{0s} is described, as before, by relationships (93) at

$$Z_{0s} = \sigma_{0s} p^{-1} B_s^{-1} (A_s K_{0s} - Y_c - \alpha \tilde{x}_s) \qquad (s = 2, \ldots, n) \tag{100}$$

In order to form a formal frequency equation, assume the boundary conditions $N_{00} = N_{0n} = 0$ ($n = s_{max}$); in addition, take $K_0 = 1$. In order to perform transition from the cross section $s = 0$ to the cross section $s = n$ using recurrent relationships (93) it is necessary first to determine two parameters Y_c and α. For this purpose we will use the condition of equilibrium (in the amplitude values) of the driven member. Using the formulae of Table 7 and (100), it can be shown that

$$\left.\begin{aligned} m_0 p^2 Y_c &= -\sum_{s=2}^{n} B_s^{-1}(K_{0s} - D_s K_{1s}) \\ J p^2 \alpha &= -\sum_{s=2}^{n} B_s^{-1} \tilde{x}_s (K_{0s} - D_s K_{1s}) \end{aligned}\right\} \tag{101}$$

In system (101) the stroboscopic modes K_{0s} and K_{1s} also depend on Y_c and α; for K_{1s} this relationship is described by Eq. 99. In order to reveal the influence of Y_c and α on K_{0s} we will use again the numerical experiment procedure. Let us introduce the following functions as linear combinations of K_{00}, Y_c and α:

$$\left.\begin{aligned} \sum_{s=2}^{n} K_{0s} B_s^{-1} &= S_{10}(p) K_{00} + S_{11}(p) Y_c + S_{12}(p) \alpha \\ \sum_{s=2}^{n} K_{0s} B_s^{-1} \tilde{x}_s &= S_{20}(p) K_{00} + S_{21}(p) Y_c + S_{22}(p) \alpha \end{aligned}\right\} \tag{102}$$

At the fixed p for the main shaft ($m = 0$), using recurrent relationships (93) and (100), first we find the value K_{0s} at the following combinations of the initial data:

1) $K_{00}^{(0)}=1,\ Y_{\mathrm{c}}^{(0)}=0,\ \alpha^{(0)}=0$ (here $Z_{0s}^{(0)}=\sigma_{0s}p^{-1}B_s^{-1}A_sK_{0s}^{(0)}$)

2) $K_{00}^{(1)}=\alpha^{(1)}=0,\ Y_{\mathrm{c}}^{(1)}=1$; 3) $K_{00}^{(2)}=0,\ Y_{\mathrm{c}}^{(2)}=0,\ \alpha^{(2)}=1$

(here and below, the number of the computation run is marked by the superscript).

It follows from (102) that

$$S_{1k}=\sum_{s=2}^{n}K_{0s}^{(k)}B_s^{-1},\quad S_{2k}=\sum_{s=2}^{n}K_{0s}^{(k)}B_s^{-1}\tilde{x}_s$$

(k is the number of the computation). Now the system of equations of Y_{c} and α may be written in the final form:

$$\left.\begin{aligned}\left(\sum_{s=2}^{n}B_s^{-1}D_s-m_0p^2-S_{11}\right)Y_{\mathrm{c}}+\left(\sum_{s=2}^{n}B_s^{-1}D_s\tilde{x}_s-S_{12}\right)\alpha=S_{10}\\ \left(\sum_{s=2}^{n}B_s^{-1}D_s\tilde{x}_s-S_{21}\right)Y_{\mathrm{c}}+\left(\sum_{s=2}^{n}B_s^{-1}D_s\tilde{x}_s^2-Jp^2+S_{22}\right)\alpha=S_{20}\end{aligned}\right\}\quad(103)$$

After determination of the roots of the set of linear algebraic equations (103), the problem is basically solved since it is reduced to the previous one (see §15). In particular, the formal frequency equation may be written as $U(p) = N_{0n}(p) = 0$ or $H_{\mathrm{m}}(p) = H_{\mathrm{d}}(p)$.

§17. REGULAR SYSTEMS OF RING STRUCTURE

General information on regular vibratory systems made of repeating blocks of the mechanisms and, associated with them, elements of the driving and driven members, were presented in §9. The analytical technique of similar systems, presented before, will be extrapolated below for the ring structure systems. We will limit ourselves only to the frequency and modal analysis on the basis of the "global" model since the obtained information allows for switching to the local model using the relationships presented in §15.

Schematization of the driven member as a system with distributed parameters. Refer to the model consisting of the driving mechanism ($s = n + 1$) and two torsional subsystems with distributed parameters ($m = 0; 1$), connected between themselves by k identical mechanisms $s = 1, ..., n$ (Fig. 24).

As before, each of the mechanisms, as well as the drive, are represented as a series chain of the discrete elements sj (s corresponds to the number of the mechanism, j corresponds to the ordinal number in the chain s). The segment s of the subsystems is limited by the cross sections in which the members of the mechanisms $s - 1$ and s are located.

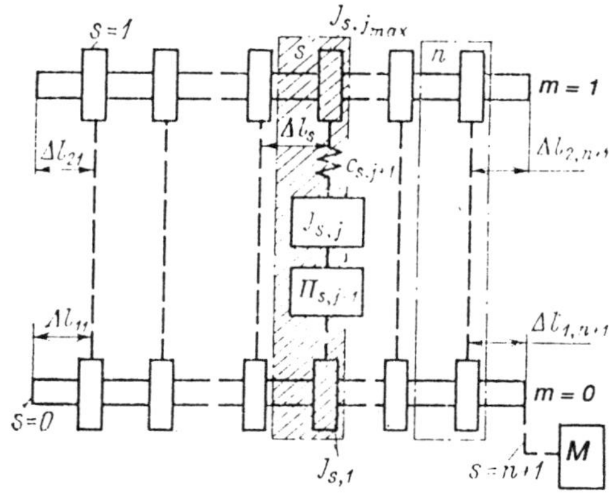

Figure 24. The dynamic model of the regular system of the ring structure.

As it has been shown in §15, the mode of vibrations in the arbitrary segment s of the subsystem m is described by relationship (92), in which the functions K_{ms} and N_{ms} are determined by recurrent relationships (93). Making these relationships specific for the considered model using the formulae for Z_{ms} from Table 7 ($v_{ms} = 3$), and taking for convenience that $K_{0s} = K_s$, $N_{0s} = N_s$, $K_{1s} = L_s$, and $N_{1s} = M_s$, obtain

$$\left.\begin{aligned} K_s &= K_{s-1}\cos\theta_{0s} + N_{s-1}\sin\theta_{0s} \\ N_s &= -K_{s-1}\sin\theta_{0s} + N_{s-1}\cos\theta_{0s} - \mu_{00}^{(s)}K_s - \mu_{01}^{(s)}L_s \\ L_s &= L_{s-1}\cos\theta_{1s} + M_{s-1}\sin\theta_{1s} \\ M_s &= -L_{s-1}\sin\theta_{1s} + M_{s-1}\cos\theta_{1s} - \mu_{10}^{(s)}K_s - \mu_{11}^{(s)}L_s \end{aligned}\right\} \quad (104)$$

Here, the functions $\mu_{m0}^{(s)}$ and $\mu_{m1}^{(s)}$ reflect the influence of components of amplitude reactive torques acting from the side of the mechanism s on the subsystem m, with $\mu_{00}^{(s)} = -\sigma_0 A_s/(pB_s)$; $\mu_{01}^{(s)} = \sigma_0/(pB_s)$; $\mu_{10}^{(s)} = \sigma_1/(pB_s)$; $\mu_{11}^{(s)} = -\sigma_1 D_s/(pB_s)$; A_s, B_s, C_s, D_s are elements of the transition matrix of the mechanism s (see above).

In the general case, as it has been shown in §15, $\theta_{ms} = p\Delta l_s/g_{ms}$. In the following discussion, assume that the moments of inertia of parts installed in the subsystems $m = 0$ and $m = 1$ are considered in the mechanism chains. In this case, $g_{ms} = \sqrt{G/\rho^*}$ where ρ^* is the density of the material; G is shear modulus, such that $\theta_{0s} = \theta_{1s} = p\Delta l_s \times \sqrt{\rho^*/G} = \theta_s$. It means that function θ_{ms} does not depend on the number of the subsystem. It should be remembered that the functions K, N, L, M in each of the considered segments simultaneously determine the amplitude torques $Q_{ms} = GI_m\, \partial X_{ms}/\partial x_{ms}$.

Suppose that the lengths of the segments Δl_s at $s = 2, ..., n$ are equal between themselves, and the mechanisms are fully identical. In this case, excluding the driving mechanism $s = n + 1$ and the outer segments $s = 1$, $s = n + 1$, the system becomes regular (the repetitive block is cross-hatched in Fig. 24). Further, considering (93) at $s = 2, ..., n$ as a system of differential equations, we will seek solution in the form $K_s = \lambda K_{s-1}$, $N_s = \lambda N_{s-1}$, $L_s = \lambda L_{s-1}$, $M_s = \lambda M_{s-1}$. After substitution into (93), we have

$$\left.\begin{aligned}
&K_{s-1}(\cos\theta-\lambda)+N_{s-1}\sin\theta=0\\
&-K_{s-1}(\mu_{00}\lambda+\sin\theta)+N_{s-1}(\cos\theta-\lambda)-\mu_{01}\lambda L_{s-1}=0\\
&L_{s-1}(\cos\theta-\lambda)+M_{s-1}\sin\theta=0\\
&K_{s-1}\mu_{10}\lambda-L_{s-1}(\mu_{11}\lambda+\sin\theta)+M_{s-1}(\cos\theta-\lambda)=0
\end{aligned}\right\}$$

Excluding the trivial zero solution, we will require that the determinant of this system be zero. This leads to the following full algebraic equation of the fourth order, which appears to be symmetrical for the characteristic number λ

$$\lambda^4+a\lambda^3+b\lambda^2+a\lambda+1=0 \tag{105}$$

Here, $a=(\mu_{00}+\mu_{11})\sin\theta-4\cos\theta$; $b=(\mu_{00}\mu_{11}-\mu_{01}\mu_{10})\sin^2\theta-(\mu_{00}+\mu_{11})\times\sin2\theta+4\cos^2\theta+2$ (the subscript s at the same parameters of the regular part of the system is omitted).

The solution of the symmetrical equation (105) is sought using the known substitution $\lambda+\lambda^{-1}=z$. It transforms Eq. (105) into two quadratic equations whose roots are

$$\lambda_{1,2,3,4}=0.5(z_{1,2}\mp\sqrt{z_{1,2}^2-4}) \tag{106}$$

where $z_{1,2}=0.5\,[-a\pm\sqrt{a^2-4\,(b-2)}]$.

It is possible to show that $F=a^2-4(b-2)=[(\mu_{00}-\mu_{11})\sin\theta+2\cos\theta]^2+4\mu_{01}\mu_{10}\sin^2\theta$. Since $\mu_{01}\mu_{10}=\sigma_1\sigma_0(pB)^2>0$, then $F>0$. Consequently, z is the real number at any value of the "natural" frequency p. On the basis of (106), at $|z_{1,2}|<2$ we have a pair of conjugated roots λ. This case in the coordinate system ab (Fig. 25) is represented by region $1a$, limited by the straight lines $b=$

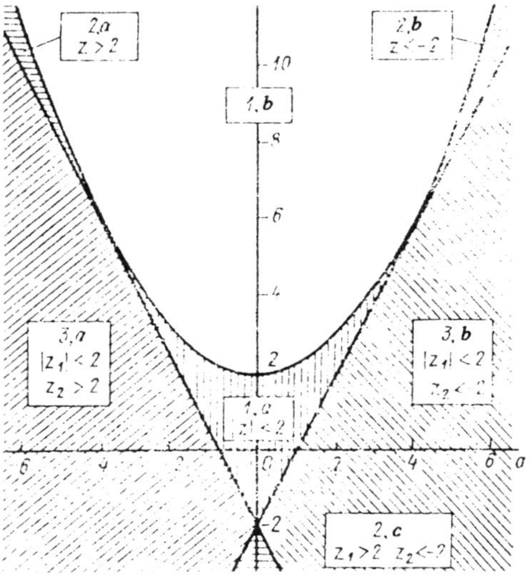

Figure 25. Ranges of existence of characteristic numbers.

$-2(1 \pm a)$ and by the parabola $b = 0.25a^2 + 2$. The straight lines intersect in the point $a = 0$, $b = -2$, and touch the parabola in the points $a = \pm 4$, $b = 6$. In region *1b* $F < 0$, which means the absence of solutions. In the regions *2a* and *2b* λ are real numbers; in the regions *3a* and *3b* one pair of roots λ is the complex conjugate numbers, and the other one is the real numbers.

Introduce now functions γ_k, f_{ks} and υ_{ks}, determined for each of the regions by the formulae presented in Table 8, and present the solution of the system of the uniform (homogeneous) differential equations as

$$\left.\begin{aligned}
K_s &= \sum_{k=1}^{2} (h_{k1} f_{ks} + h_{k2} v_{ks}) \\
N_s &= \sin^{-1}\theta \sum_{k=1}^{2} [h_{k1}(f_{k,\,s+1} - f_{ks}\cos\theta) + h_{k2}(v_{k,\,s+1} - v_{ks}\cos\theta)] \\
L_s &= \sum_{k=1}^{2} \beta_k (h_{k1} f_{ks} + h_{k2} v_{ks}) \\
M_s &= \sin^{-1}\theta \sum_{k=1}^{2} \beta_k [h_{k1}(f_{k,\,s+1} - f_{ks}\cos\theta) + h_{k2}(v_{k,\,s+1} - v_{ks}\cos\theta)]
\end{aligned}\right\} \qquad (107)$$

where $\beta_k = -\mu_{01}^{-1}\sin^{-1}\theta(z_k + \mu_{00}\sin\theta - 2\cos\theta)$; $s = 1, \ldots, n$.

The functions K_s and M_s, as before (see §15), are describing the stroboscopic mode of vibrations corresponding to the cross sections of the input and output members of the actuators. For the amplitude values of loads in these cross sections we have $Q_{0s}(0) = p\sigma_0^{-1}N_s$ and $Q_{1s}(0) = p\sigma_1^{-1}M_s$, therefore the functions N and M_s characterize the "stroboscopic" distribution of the load at free vibrations for the subsystems $m = 0$ and $m = 1$.

Determine now the functions h_{k1} and h_{k2} from the boundary conditions $N_0 = 0$, $M_0 = 0$, and $M_{n+1} = 0$, which follow from the absence of loads in these cross sections. In addition, assume the modal coefficient in the initial cross section of the system $m = 0$ to be $K_0 = 0$. Using these conditions and the recurrent relationships (104), we find the functions K, N, L, M on the boundaries of the regular part of the system, i.e., at $s = 1$ and $s = n$. On the other hand, these functions can be determined by the direct substitution of these values into relationships (107). Combining these two results we obtain the following system of linear algebraic equations relative to the functions h_{k1} and h_{k2} ($k = 1, 2$):

$$\left.\begin{aligned}
& h_{11} + h_{21} = \cos\theta_{01} \\
& \sum_{k=1}^{2} [h_{k1}(f_{k2} - \cos\theta) + h_{k2} v_{k2}] = -\sin\theta_{01}\sin\theta \\
& \sum_{k=1}^{2} \beta_k [h_{k1}(f_{k2} - \cos\theta + \sin\theta\,\mathrm{tg}\,\theta_{11}) + h_{k2} v_{k2}] = 0 \\
& \sum_{k=1}^{2} \beta_k \{h_{k1}[f_{k,\,n+1} - f_{kn}(\cos\theta + \sin\theta\,\mathrm{tg}\,\theta_{1,\,n+1})] \\
& \quad + h_{k2}[v_{k,\,n+1} - v_{kn}(\cos\theta + \sin\theta\,\mathrm{tg}\,\theta_{1,\,n+1})]\} = 0
\end{aligned}\right\} \qquad (108)$$

Table 8. Functions $\lambda, \gamma_k, f_{ks}, \upsilon_{ks}$.

Range	k	z_k	λ	γ_k	f_{ks}	υ_{ks}
1, *a*	1; 2	$\vert z_k \vert < 2$	$\vert \lambda_k \vert e^{\pm i\gamma_k}$	$\arccos 0.5z_k$	$\cos(s-1)\gamma_k$	$\sin(s-1)\gamma_k$
2, *a*	1; 2	$z_k > 2$	$\vert \lambda_k \vert e^{\pm\gamma_k}$	Arch $0.5z_k$	$\mathrm{ch}(s-1)\gamma_k$	$\mathrm{sh}(s-1)\gamma_k$
2, *b*	1; 2	$z_k < -2$	$\vert \lambda_k \vert e^{\pm\gamma_k}$	$\gamma_k^0 + i\pi$, where $\gamma_k^0 =$ Arch $0.5\vert z_k \vert$	$(-1)^{s-1}\mathrm{ch}(s-1)\gamma_k^0$	$(-1)^{s-1}\mathrm{sh}(s-1)\gamma_k^0$
2, *c*	1	$z_1 > 2$	$\vert \lambda_k \vert e^{\pm\gamma_k}$	Arch $0.5z_1$	$\mathrm{ch}(s-1)\gamma_1$	$\mathrm{sh}\,(s-1)\gamma_1$
	2	$z_2 < -2$	$\vert \lambda_k \vert e^{\pm\gamma_k}$	$\gamma_2^0 + i\pi$	$(-1)^{s-1}\mathrm{ch}(s-1)\gamma_2^0$	$(-1)^{s-1}\mathrm{sh}(s-1)\gamma_2^0$
3, *a*	1	$\vert z_1 \vert < 2$	$\vert \lambda_1 \vert e^{\pm i\gamma_1}$	$\arccos 0.5z_1$	$\cos(s-1)\gamma_1$	$\sin(s-1)\gamma_1$
	2	$z_2 > 2$	$\vert \lambda_2 \vert e^{\pm\gamma_2}$	Arch $0,5z_2$	$\mathrm{ch}(s-1)\gamma_2$	$\mathrm{sh}(s-1)\gamma_2$
3, *b*	1	$\vert z_1 \vert < 2$	$\vert \lambda_1 \vert e^{\pm i\gamma_1}$	$\arccos 0.5z_1$	$\cos(s-1)\gamma_1$	$\sin(s-1)\gamma_1$
	2	$z_2 < -2$	$\vert \lambda_2 \vert e^{\pm\gamma_2}$	$\gamma_2^0 + i\pi$	$(-1)^{s-1}\mathrm{ch}(s-1)\gamma_2^0$	$(-1)^{s-1}\mathrm{sh}(s-1)\gamma_2^0$

Having solved this system for the fixed value p, we obtain $h_{11}(p)$, $h_{12}(p)$, $h_{21}(p)$, $h_{22}(p)$. In addition to the boundary conditions presented above, we have one more condition, namely, the equality of amplitude loads in the place of the attachment of the operating mechanism ($s = n + 1$). Having used this condition, we obtain the formal frequency equation in the following form:

$$\Phi(p) = H_{\mathrm{m}}(p) - H_{\mathrm{d}}(p) = 0$$

Here, $H_{\mathrm{m}}(p) = p\sqrt{J_0/c_0}(K_n \operatorname{tg}\theta_{0,\,n+1} - N_n)/(K_n + N_n \operatorname{tg}\theta_{0,\,n+1})$; $H_{\mathrm{d}}(p) = R_{n+1}/c_0$, where $c_0 = \sqrt{GI_0/l}$, J_0 are the torsional stiffness and the moment of inertia of the distributing shaft ($m = 0$), respectively; $R_{n+1} = A_{n+1}/B_{n+1}$ for the fixed end of the driving mechanism and $R_{n+1} = C_{n+1}/D_{n+1}$ for the free end of the driving mechanism (counting from the shaft); the functions K_n and N_n are determined by formulae (107) at $s = n$ and at the values h_{k1}, h_{k2}, which were found at the solution of the system (108).

The "natural" frequencies p can be found either as the points of the intersection of the curves of functions $H_{\mathrm{m}}(p)$ and $H_{\mathrm{d}}(p)$ or on a computer.

Nonstationary modes of vibrations are determined for each root p_r of the formal frequency equation by relationships (92) and (96), in which functions K_s, N_s, L_s, M_s should be calculated using formulae (107) at values $h_{k1}(p)$ and $h_{k2}(p)$ ($k = 1, 2$) obtained from solving system (108).

The procedure presented above may be used also when there are some violations of the regularity conditions, for example, if the values Δl_s are not strictly identical but slightly fluctuate about the average value.

Modelling of the driven member as a rigid body with programmed translational movement. The general format of such a model has been analyzed above in §16. Here, we consider a particular case when all the mechanisms are equally spaced at Δl from one another (Fig. 26), and the driving mechanism is attached at the one end of the distributing shaft, e.g., at the cross section $s = 0$. The system is the regular one with the exception of the end segments [22].

The recurrent relationships for the functions K_s and N_s coincide fully with the first two equations of system (104). Instead of the second pair of equations

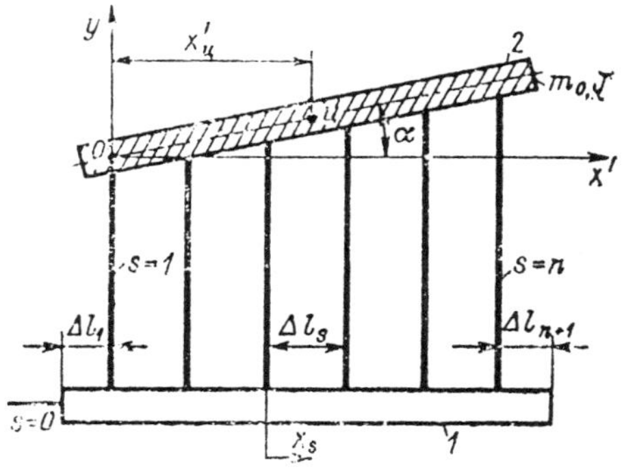

Figure 26. The analytical model of the drive for translational programmed motion of the driving member: 1) distributing shaft; 2) working member; $s = 0$ is the driving mechanism; $s = 1, \ldots, n$ are transmission mechanisms.

which correspond to the driven member, we write the obvious geometrical relationship

$$L_s = L_1 + \alpha(s-1)\Delta l \tag{109}$$

Here, L_1 is the amplitude of the translational displacement of the cross section $s = 1$, and α is the angular amplitude of the driven member. First, establish a correlation between L_1, α, and the functions K_s, N_s. The recurrent relationships mentioned above, subject to (109), represent a system of heterogeneous differential equations whose solution can be reduced to:

$$\begin{aligned} K_s &= h_1 f_s + h_2 v_s - \chi_1 L_1 - \chi_2 \alpha(s-1) \\ N_s &= h_1(f_{s+1} - f_s \cos\theta) + h_2(v_{s+1} - v_s \cos\theta) \\ &+ \chi_1 L_1(1-\cos\theta) + \chi_2 \alpha[s-(s-1)\cos\theta] \end{aligned} \tag{110}$$

where $\chi_1 = -\mu_{01}/(2\mathrm{tg}\,0.5\theta + \mu_{00})$; $\chi_2 = \chi_1 \Delta l$; $\mu_{00} = -\sigma_0 A/(pB)$; $\mu_{01} = \sigma_0/(pB)$.

The functions f_s and v_s are determined depending on the value $\varkappa = \cos\theta - 0.5\mu_{00}\sin\theta$. At $|\varkappa| < 1$ we have $f_s = \cos(s-1)\gamma$ and $v_s = \sin(s-1)\gamma$ where $\gamma = \arccos\varkappa$; at $\varkappa > 1$, $f_s = \mathrm{ch}(s-1)\gamma$ and $v_s = \mathrm{sh}(s-1)\gamma$ where $\gamma = \mathrm{Arch}\,\varkappa$; at $\varkappa < -1$, $f_s = (-1)^{s-1}\mathrm{ch}(s-1)\gamma^0$ and $v_s = (-1)^{s-1}\mathrm{sh}(s-1)\gamma^0$ where $\gamma^0 = \mathrm{Arch}\,|\varkappa|$.

The parameters L_1, α and h_1, h_2 are connected by the equilibrium equations of the rigid body, which for the amplitude values of forces and torques can be reduced to:

$$\left.\begin{aligned} u_{11}L_1 + u_{12}\alpha &= \mu_{10}(S_{11}h_1 + S_{12}h_2) \\ u_{21}L_1 + u_{22}\alpha &= \mu_{10}\Delta l(S_{21}h_1 + S_{22}h_2) \end{aligned}\right\} \tag{111}$$

where

$u_{11} = -[m_0 p^2 + k(\mu_{10}\chi_1 + \mu_{11})]$; $u_{12} = -[0.5k(k-1)(\chi_2\mu_{10} + \mu_{11}\Delta l) + m_0 p^2 x_c']$; $u_{21} = -0.5k(k-1)\Delta l(\chi_1\mu_{10} + \mu_{11})$; $u_{22} = -[Jp^2 + \Delta l^2 k(k-1)(2k-1)(\mu_{11} + \mu_{10}\chi_2)/6]$; $S_{11} = \sum_{s=1}^{k} f_s$; $S_{12} = \sum_{s=1}^{k} v_s$; $S_{21} = \sum_{s=1}^{k}(s-1)f_s$; $S_{22} = \sum_{s=1}^{k}(s-1)v_s$; $J = J_0 + m_0 x_c'^2$; $\mu_{10} = B^{-1}$; $\mu_{11} = -D/B$.

Here, m_0 is the mass of the driven member; J_0 is the moment of inertia of the driven member about the axis passing through the center of mass; p is the "natural" frequency; k is the number of identical mechanisms.

After solution of system (111) we obtain

$$\begin{bmatrix} L_1 \\ \alpha \end{bmatrix} = \begin{bmatrix} \varkappa_{11} & \varkappa_{12} \\ \varkappa_{21} & \varkappa_{22} \end{bmatrix} \begin{bmatrix} h_1 \\ h_2 \end{bmatrix}$$

where $\varkappa_{11} = \Delta_*^{-1}\mu_{10}(S_{11}u_{22} - \Delta l S_{21}u_{12})$; $\varkappa_{12} = \Delta_*^{-1}\mu_{10}(S_{12}u_{22} - \Delta l S_{22}u_{12})$;

$\varkappa_{21}=\Delta_*^{-1}\mu_{10}(u_{11}\Delta l S_{21}-u_{21}S_{11})$; $\varkappa_{22}=\Delta_*^{-1}\mu_{10}(u_{11}\Delta l S_{22}-u_{21}S_{12})$; $\Delta_* = u_{11}u_{22}-u_{12}u_{21}$.

Then, after substitution of L_1 and α into relationships (110) we have

$$K_s=\delta_{11}^{(s)}h_1+\delta_{12}^{(s)}h_2;\ N_s=\delta_{21}^{(s)}h_1+\delta_{22}^{(s)}h_2,\ \text{where}\ \delta_{11}^{(s)}=f_s+\chi_1\varkappa_{11}+\chi_2\varkappa_{21}(s-1);\ \delta_{12}^{(s)}=v_s+\chi_1\varkappa_{12}+\chi_2\varkappa_{22}(s-1);\ \delta_{21}^{(s)}=\sin^{-1}\theta[f_{s+1}-f_s\cos\theta+\chi_1\varkappa_{11}(1-\cos\theta)+\chi_2\varkappa_{21}[s-(s-1)\cos\theta)];\ \delta_{22}^{(s)}=\sin^{-1}\theta[v_{s+1}-v_s\cos\theta+\chi_1\varkappa_{12}(1-\cos\theta)+\chi_2\varkappa_{22}[s-(s-1)\cos\theta)]$$

The coefficients h_1 and h_2 can be determined using the boundary condition N_{n-1} at $K_1 = 1$. Hence

$$h_1=(\delta_{11}^{(1)}+\delta_{12}^{(1)}\xi)^{-1}$$

$$h_2=\xi(\delta_{11}^{(1)}+\delta_{12}^{(1)}\xi)^{-1}$$

where $\xi = -(\delta_{21}^{(n)}-\delta_{11}^{(n)}\operatorname{tg}\theta_{n+1}):(\delta_{22}^{(n)}-\delta_{12}^{(n)}\operatorname{tg}\theta_{n+1})$.

In addition, we have the equality condition for amplitude moments in the point of attachment of the drive. This condition is reduced to the format of the formal frequency equation (81) at

$$H_d(p) = A_d l/(B_d G I_0)\ \text{(the end is fixed)}$$

$$H_d(p) = C_d l/(D_d G I_0)\ \text{(the end is free)}$$

$$H_M(p)=[\sin\theta_1+\cos\theta_1(U+\mu_{00})]\sum_{s=1}^{n+1}\theta_s/[\cos\theta_1-\sin\theta_1(U+\mu_{00})]$$

where $U=[\mu_{01}(\varkappa_{11}+\xi\varkappa_{12})+\delta_{21}^{(1)}+\xi\delta_{22}^{(1)}]/(\delta_{11}^{(1)}+\xi\delta_{12}^{(1)})$; A_d, B_d, C_d, D_d are the elements of the transition matrix of the driving mechanism.

The curves $H_m(p)$ and $H_d(p)$ for a typical case are presented in Fig. 27*a*. In Fig. 27*b* a number of modes of vibration are shown which correspond to the "natural" frequencies determined as the intersection points of the curves $H_m(p)$ and $H_d(p)$. At $r = 1$ the mode of vibration does not have nodal points; at $r = 2$ the driving and driven members of each mechanism vibrate in phase, while the outer mechanisms between themselves vibrate in antiphase; at $r = 3$ the driving and driven members vibrate in phase; at $r = 4$ both the outer mechanisms and the driving and driven members of each mechanism vibrate in antiphase. It is of interest that at the increased stiffness of the driving mechanism, the curve $H_d(p)$

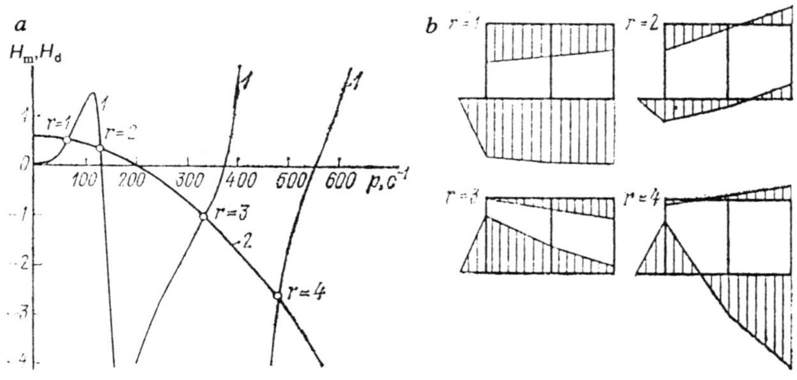

Figure 27. Curves $H_m(p)$, $H_d(p)$, and the modes of vibrations.

may pass above the maximum of the function $H_m(p)$; in this case the first mode would have the shape corresponding to $r = 3$ at the initial (reduced) stiffness.

§18. SPECIFICS OF SYMMETRIC REGULAR SYSTEMS

Consider now a dynamic model of the drive consisting of the operating member (OM) and k identical actuators. The example of such a model is shown in Fig. 28. Suppose that the drive may be considered as absolutely rigid, which for the frequency analysis is equivalent to clamping of the corresponding element of the mechanism; such idealization is frequently useful at very small magnitudes of Π' when OM, together with actuators, represents an isolated vibratory system whose frequency and modes, practically, do not depend on the drive characteristics. The operating member can be modelled as a torsional subsystem of two types: with lumped and distributed parameters.

This model could be analyzed as a particular case of the model of the ring structure considered in §15. However, it is much easier to use the results of the analysis of the regular branched system from §10 if the distributing shaft is replaced by the operating member and the ends of the mechanisms are regarded as not free, but fixed. One should remember that in this case it is necessary to count the elements from the operating member; it is especially important for kinematic characteristics in the transition matrices (Π' should be replaced by $1/\Pi'$).

Using the apparatus of the differential equations presented in §§9 and 10, consider again the criterion $\varkappa_s$ determining the shape of the stroboscopic mode of vibrations

$$\varkappa_s = 0.5\,(A_{0s} + D_{0s} + R_s B_{0s})$$

In this case, due to immobility of the ends of the mechanisms, the function R_s, which is proportional to the amplitude of the reactive torque acting from the mechanism s to the operating member, is equal to $R_s = A_s / B_s$. It is possible to show that, unlike the models considered in §10, in this symmetrical case always $|\varkappa_s| < 1$. Consequently, the stroboscopic mode of vibration and load is as follows

$$\left.\begin{aligned} a_s &= K_s = h_1 \cos s\gamma + h_2 \sin s\gamma \\ Q_s &= (a_{s+1} - a_s A_{0s}) B_{0s}^{-1} \end{aligned}\right\} \quad (112)$$

where $\gamma = \arccos \varkappa_s$ h_1, h_2 are functions determinable using the boundary conditions.

As it has been shown in §10, the criterion $\varkappa_s$ and, consequently, γ, do not depend on the attachment point location inside the segment s of the kinematic branch.

Modelling of the operating member as a subsystem with lumped parameters. The operating member is now presented as a series connection of similar discs located on the places of the attachment of mechanisms. The discs are connected by elastic elements with stiffness coefficient Δc_0 corresponding to the torsional stiffness of the segment of the operating member. Thus, inertial properties of the operating member are directly included into the model of the actuator.

If the vibratory system of the actuator has H_1 degrees of freedom, then the full number of degrees of freedom of the model is $H = kH_1$. In this case $A_{0s} = D_{0s} = 1$, $B_{0s} = 1/\Delta c_0$, and, consequently, $\varkappa = \cos \gamma = 1 + 0.5R/\Delta c_0$ (here and below, at the identical parameters subscript s is omitted).

Using relationships (112) and the boundary conditions $Q_0 = 0$, $Q_k = 0$, the following equation results (see Table 4)

$$\cos(k - 0.5)\gamma = \cos(k + 0.5)\gamma$$

whose solution is $\gamma = \pi j/k$ where $j = 0, 1, ..., k - 1$.

Hence, the formal frequency equation may be written as:

$$1 + 0.5R/\Delta c_0 = \cos(\pi j/k) \quad (113)$$

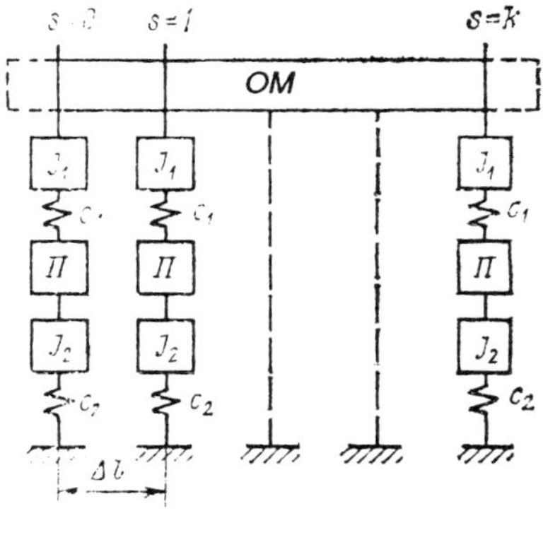

Figure 28. The dynamic model of the symmetric regular system.

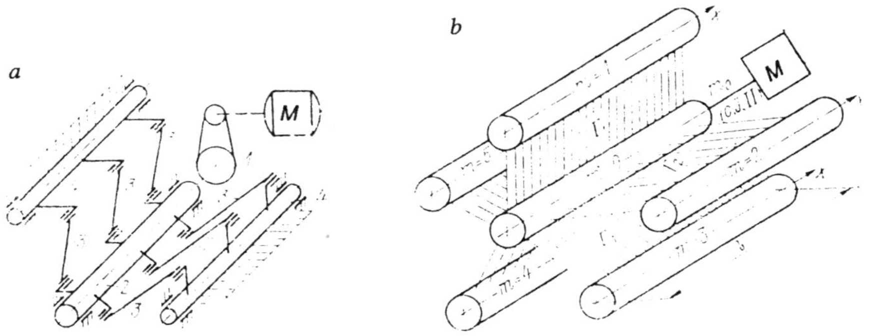

Figure 29. Multicontour ring structure drive and its dynamic model.

Below, Eq. (113) is specified for the case when each of the mechanisms is represented by the vibratory system with two degrees of freedom (see Fig. 29), which is represented by the following transition matrix:

$$\Gamma = \begin{bmatrix} 1 & c_2^{-1} \\ 0 & 1 \end{bmatrix} \begin{bmatrix} 1 & 0 \\ -J_2 p^2 & 1 \end{bmatrix} \begin{bmatrix} (\Pi')^{-1} & 0 \\ 0 & \Pi' \end{bmatrix} \begin{bmatrix} 1 & c_1^{-1} \\ 0 & 1 \end{bmatrix} \begin{bmatrix} 1 & 0 \\ -J_1 p^2 & 1 \end{bmatrix}$$

(see above about handling the function Π').

Having multiplied the matrices, we find $R = A/B$, thus the equation becomes:

$$J_1 J_2 p^4 - \{J_1(c_2 + c_1 \Pi'^2) + J_2[c_1 + 2\Delta c_0(1 - \cos j\pi/k)]\} p^2 + c_1 c_2 + 2\Delta c_0(c_2 + c_1 \Pi'^2)(1 - \cos j\pi/k) = 0 \quad (114)$$

$$(j = 0, 1, \ldots, k-1)$$

It means that in order to determine $2k$ "natural" frequencies, the biquadratic equation (114) should be solved k times. At $J_2 = 0$ we have the vibratory system with k degrees of freedom, and

$$p_r = \sqrt{[c_* + 2\Delta c_0(1 - \cos j\pi/k)]/J_1} \quad (115)$$

where $c_* = c_1 c_2/(c_2 + c_1 \Pi'^2)$, $r = j + 1$.

The analysis of formula (115) shows that at $\Delta c_0 \ll c_*$, the spectrum of the "natural" frequencies will have a rather high density in the frequency range

$$\sqrt{c_* J_1} \leq p_r < \sqrt{(c_* + 4\Delta c_0) J_1}$$

Attention should be paid to the fact that, irrespective of the number of the mechanisms, *the lowest "natural" frequency of the whole system is equal to the partial frequency of the actuator taken separately*. A similar conclusion may be made at any number of degrees of freedom of the subsystem of the actuator. *The combination of variability of the "natural" frequencies, which is specific to reonomic vibratory systems, and stationarity of vibratory modes characteristic to the systems with constant parameters is a rather specific characteristic property of the considered model.* Actually, the mode of vibrations determined by the first of relationships (112) subject to the boundary conditions can be expressed as

$$a_{s}^{(r)} = \cos[s\pi(r-1)/k] + (-1)^{r-1}\mathrm{tg}[\pi(r-1)/(2k)]\sin[s\pi(r-1)/k]$$

Since $\gamma_r = \pi(r-1)/k$ does not depend on time, the mode of vibrations is stationary. At the same time, the natural frequency p can change in time; it is easy to see by analyzing, for example, formula (115). From this formula, at $\Pi' \neq$ const we have $c_* = c_*(\tau)$ and $p_r = p_r(\tau) \neq$ const. The determining role in such a nontrivial behavior plays the full dynamic symmetry of the considered regular system.

Modelling of the operating member as a subsystem with distributed parameters. For the system to remain strictly regular in this case, the ends of the operating member should be increased by segments $0.5\Delta l$ long, as shown in Fig. 29, by hatching. Using the formulae from Table 2, at $H_m = 0$ obtain $\gamma_r = \pi(r-1)/k$ ($r = 1, ..., k$). Here, the formal frequency equation is

$$R(p_r) = 2[\cos\gamma_r - \cos\theta(p_r)]\,\Delta c_0\theta(p_r)/\sin\theta(p_r) \tag{116}$$

This equation, unlike the previous case, is transcendental. It should be remembered that the function $R(p)$ from another side is determined as $R = A/B$. In particular, for the case considered above when the mechanism was presented as a vibratory system with two degrees of freedom,

$$R(p) = \frac{J_1J_2p^4 - [J_2c_1 + J_1(c_2 + c_1\Pi'^2)]p^2 + c_1c_2}{c_2 + c_1\Pi'^2 - J_2p^2}$$

At $J_2 = 0$ we have $R(p) = c_* - J_1p^2$ where $c_* = c_1c_2/(c_2 + c_1\Pi'^2)$. At $\theta \to 0$ formulae (113) and (116) coincide. In engineering calculations discrepancy between these two formulae are small at $p < 0.5g_0/\Delta l$ where $g_0 = \sqrt{GI/\rho}$.

The stroboscopic mode of vibration in this case also retains its stationary properties in spite of variance of the "natural" frequencies.

We may use the condition of dynamic symmetry in such cases when the system is not strictly regular as, for example, at the absence of the segments of operating members shown in Fig. 28 by hatching. Assuming in the formulae of Table 2 $\theta_1 = 0$, $\theta_{n+1} = 0$, and substituting k into $k_1 = k - 1$, the formal frequency equation becomes $H_m = H_d$, where in our case $H_d = \sigma R/p$ (see §7). Omitting transformations, we obtain

$$\cos k_1\gamma(p) = \pm 1 - 0.5R(p)\sin\theta(p)/(\theta(p)\Delta c_0)$$

where $\cos\gamma = \cos\theta + 0.5R\sin\theta/(\Delta c_0\theta)$; R (see above).

At $\sin\theta/\theta \approx 1$ we obtain $\gamma_r(p_r) = \pi(r-1)/(k_1+1) = \pi(r-1)/k$ ($r = 1, \ldots, k$), which coincides with earlier results.

In addition to this procedure, for solving analogous problems we can use apparatus of the group theory and the perturbation theory [3].

§19. ANALYSIS OF SOME CONTINUAL DYNAMIC MODELS

Mathematical model. Let's analyze a drive in which the common distributing shaft is connected with m subsystems of the ring structure consisting of transmission mechanisms and operating members (Fig. 29a) [17, 21]. Such a structure, for example, is used in modern warp-knitting machines in which the number of the subsystems is $m_* = 4$. In Fig. 29 we present a generalized dynamic model of the drive, which consists of the carrier subsystem $m = 0$ ("trunk") and the subsystems $m = 1, ..., m_*$. The carrier subsystem, imitating the main shaft, is connected with the motor M by the driving mechanism m_0, which is presented as a series chain of discrete elements—elastic (c), inertial (J), and kinematic (Π). The operating members and the main shaft are modelled as torsional subsystems with distributed parameters. In the real drive the operating members are connected with the "trunk" $m = 0$ by the finite number of mechanisms (see Fig. 24), each of which can be presented by the model with lumped parameters. However, in accordance with the method presented in §12 we use the continual representation of these mechanisms in the form of some pseudomedium formed by "spreading" of elastic, inertial, and kinematic characteristics along the axis x.

When the mechanisms are uniformly distributed along the axis x and the polar cross sectional moments of inertia I_m are constant, then for each of the subsystems the differential equation of the format (76) is valid. However, in this case instead of one equation (76) we obtain the following system of differential equations describing the nonstationary modes of vibrations:

$$\left.\begin{aligned} & X_0'' + \sum_{m \neq 0}^{m_*} P_{0m}(\tau) X_m = 0 \\ & X_m'' + P_{m0}(\tau) X_0 + P_{mm}(\tau) X_m = 0 \qquad (m = 1, \ldots, m_*) \end{aligned}\right\} \qquad (117)$$

where $P_{00} = (\rho_0 p^2 - \sum_{m=1}^{m_*} A_m B_m^{-1})/(GI_0)$; at $m \neq 0$ $P_{0m} = (B_m GI_0)^{-1}$
$P_{m0} = (B_m GI_m)^{-1}$, $P_{mm} = (\rho_m p^2 - D_m B_m^{-1})/(GI_m)$; $(\;)' = \partial/\partial x$
τ is "slow" time.

The functions A_m, B_m, D_m are the appropriate elements of the distributed modified transition matrices Γ_m for mechanisms of the subsystem m (see §12).

Method of determining "natural" frequencies and nonstationary modes of vibration. The particular solution of the system of equations (117) is sought in the form $X_m = h_m e^{\tilde{\lambda} x}$. After substitution into (117), the following equation for the characteristic index $\tilde{\lambda}$ is obtained

$$\tilde{\lambda}^2 + P_{00} - \sum_{m=1}^{m_*} P_{0m} P_{m0} / (\tilde{\lambda}^2 + P_{mm}) = 0 \qquad (118)$$

The roots of Eq. (118) can be either pure, imaginary, or real. For presenting the solution X_m in the general form, introduce the following functions

$$f_k = \begin{cases} \cos\theta_k(x) & \text{for } \tilde{\lambda}_k^2 < 0 \\ \operatorname{ch}\theta_k(x) & \text{for } \tilde{\lambda}_k^2 > 0 \end{cases} \qquad v_k = \begin{cases} \sin\theta_k(x) & \text{for } \tilde{\lambda}_k^2 < 0 \\ \operatorname{sh}\theta_k(x) & \text{for } \tilde{\lambda}_k^2 > 0 \end{cases} \tag{119}$$

where $\theta_k = |\tilde{\lambda}_k| x$.

Then

$$X_m = \sum_{k=0}^{m_*} h_{mk} [f_k(x, \tau) + \alpha_k(\tau) v_k(x, \tau)] \tag{120}$$

The functions $\alpha_k(\tau)$ define the phase of the solution; the functions $h_{mk}(\tau)$ are interconnected as components of the eigenvector

$$\beta_{mk} = h_{mk}/h_{0k} = -P_{m0}/(\tilde{\lambda}_k^2 + P_{mm}) \quad (\beta_{0k} = 1)$$

Function $\beta_{mk}(\tau)$, independent of boundary conditions, will be called the coefficient of spatial distribution. Then, the following boundary conditions are written: $X_m'(0) = 0$ ($m = 0, \ldots, m_*$); $X_m'(l) = 0$ ($m = 1, \ldots, m_*$). In addition, normalization of the modes of vibration is performed with assumption of $X_0(l) = 1$. Using relationships (119, 120) we have

$$X_m' = \sum_{k=0}^{m_*} \beta_{mk} h_{0k} |\tilde{\lambda}_k| (\partial f_k/\partial\theta_k + \alpha_k f_k)$$

Since $v_k(0) = 0$ and $f_k(0) \neq 0$, the conditions $X_m'(0) = 0$ are satisfied at $\alpha_k = 0$. The other conditions form the following system of equations relative to h_{0k}

$$\left.\begin{aligned} &\sum_{k=0}^{m_*} h_{0k} f_k^* = 1 \\ &\sum_{k=0}^{m_*} h_{0k}\beta_{mk}(\partial f/\partial\theta)^* = 0 \qquad (m = 1, \ldots, m_*) \end{aligned}\right\} \tag{121}$$

where the asterisk corresponds to the argument $x = l$.

Solving the system (121), we determine $m_* + 1$ functions h_{0k} ($k = 0, \ldots, m_*$). The unused boundary condition, maintaining the equality of the amplitude torques in the cross section $x = l$ for the subsystem of the "trunk" ($m = 0$) and the driving mechanism, serves as a formal frequency equation

$$\Phi(p) = H_m(p) - H_d(p) = 0 \tag{122}$$

where $H_m = lX_0'(l)/X_0(l) = \sum_{k=0}^{m_*} h_{0k}\theta_k^*(\partial f/\partial\theta)^*$; $H_d = -lR_n/(GI_0) = -c^{-1}R_n$ ($R_n = A_n/B_n$, c_0 is the torsional stiffness of the main shaft).

The "natural" frequencies p_r are the roots of Eq. (122). The substitution of values $p_r(\tau)$ into relationship (120) determines the nonstationary modes of vibrations $X_m^{(r)}$. The nonstationary modal coefficients for mechanisms are determined by relationships (96) at $v_{0s} = 3$; $K_{0s} = X_0(x_s)$; $K_{ms} = X_m(x_s)$, where x_s is the coordinate of the corresponding cross section of the input member of the actuator.

From analysis of Eq. (122) and the modes of vibrations, the dynamic coupling of individual subsystems can be revealed and a number of important problems of dynamics of the drive can be solved [21].

Analysis of the continual model of the ring structure on the quasi-static level. Since the systems of the ring structure are statically indeterminable, an issue of distribution of slowly changing components of the technological and kinetostatic loads, as well as deformations caused by these loads, is of interest. As an example, two subsystems connected by actuating mechanisms ($m = 0, 1$) are considered.

Suppose that a slowly changing kinetostatic load $\chi(x, \tau)$ is applied to the driven system. Then the corresponding deformation of the system γ_m can be described by equations like (75) at $\varphi = \gamma_m$; $\partial^2\gamma_m/\partial t^2 = 0$. Here,

$$-GI_0\gamma_0'' - S_0(\gamma_0, \gamma_1) = 0 \qquad -GI_1\gamma_1'' - S_1(\gamma_0, \gamma_1) = \chi \tag{123}$$

Expanding the functions S_m, which describe the distributed reactive torques acting from the mechanisms on the system m, we obtain:

$$\left.\begin{aligned} &\gamma_0'' + P_{00}\gamma_0 + P_{01}\gamma_1 = 0 \\ &\gamma_1'' + P_{10}\gamma_0 + P_{11}\gamma_1 = q \end{aligned}\right| \tag{124}$$

where $P_{00} = -A/(BGI_0)$; $P_{01} = (BGI_0)^{-1}$; $P_{10} = (BGI_1)^{-1}$; $P_{11} = -D/(BGI_1)$; $q = -\chi/(GI_1)$. At the determination of the elements of the distributed modified transition matrix $\tilde{\Gamma}$ it should be assumed that $p = 0$ (see §12).

Since $AD - CB = 1$ and $C = 0$ at $p = 0$, we have $AD = 1$. It is easy to see that $P_{00}P_{11} = P_{01}P_{10}$. It is possible to show that at the quasistatic conditions considered

$$\tilde{\lambda}_0^2 = 0; \tilde{\lambda}_1^2 = -(P_{00} + P_{11}) = (AI_1 + DI_0)/(GBI_0I_1)$$

Let's assume a uniform distribution of loads corresponding to $\chi(x) = \text{const}$ and $q(x) = \text{const}$. Considering coincidence of the exponent at the load with the double root of the characteristic equation ($q = q\exp(\tilde{\lambda}_0 x)$), solution of system (124) is

$$\left.\begin{aligned} &\gamma_0 = h_0(1 + \alpha_0 x) + h_1(\operatorname{ch}\lambda x + \alpha_1 \operatorname{sh}\lambda x) + b_{00} + b_{02}x^2 \\ &\gamma_1 = \beta_0 h_0(1 + \alpha_0 x) + \beta_1 h_1(\operatorname{ch}\lambda x + \alpha_1 \operatorname{sh}\lambda x) + b_{12}x^2 \end{aligned}\right\} \tag{125}$$

where $\beta_0 = -P_{00}/P_{01}$; $\beta_1 = -P_{11}/P_{01}$; $\lambda = \tilde{\lambda}_1$; h_0, h_1, α_0, α_1 are integration constants determinable by boundary conditions.

The coefficients b_{00}, b_{02}, b_{12} in the particular solution of the system of heterogeneous equations are found by a usual method. Here,

$$b_{00} = P_{01}q/[P_{00}(P_{00} + P_{11})] \qquad b_{02} = -0.5P_{01}q/(P_{00} + P_{11})$$

$$b_{12} = 0.5P_{00}q/(P_{00} + P_{11})$$

The boundary conditions are $X_0'(0) = 0$; $X_1'(0) = 0$; $c_0 l X_0'(l) = -c_d X_0(l)$; $X_1'(l) = 0$, where c_0 is the stiffness coefficient of the driving subsystem ($m = 0$), $c_d = A_d / B_d$. (The third condition corresponds to the equality of torques at the joint of the subsystem $m = 0$ ($x = l$) and the drive). The first two boundary conditions give $\alpha_0 = 0$ and $\alpha_1 = 0$; the last two conditions allow for determination of h_0 and h_1:

$$h_0 = -\frac{\chi l}{AI_1/I_0 + D}\left[c_0^{-1}\left(\frac{I_1 A^2 \operatorname{cth}\theta_*}{I_0 \theta_*} + 0.5\right) + c_0^{-1}\left(\frac{I_1 A^2}{I_0} + 1\right) + \frac{DB}{l}\right]$$

$$h_1 = -\frac{\chi I_1 A^2}{c_0 I_0 (AI_1/I_0 + D)\theta_* \operatorname{sh}\theta_*}$$

where $\theta_* = \lambda l$.

If the actuators are presented as a series connection of the kinematic element Π and the elastic element, then $A = \Pi'$, $B = l/(c_m \Pi')$, $C = 0$, $D = (\Pi')^{-1}$ where c_m is the overall stiffness coefficient of all the actuators reduced to the driven shaft.

In the particular case at the absolutely rigid main shaft ($I_0 \to \infty$, $c_0 = GI_0/l \to \infty$) by means of formulae (125) we obtain

$$\gamma_0 = h_0 + b_{00} = Q\Pi'/c_d \qquad \gamma_1 = Q(\Pi'^2/c_d + c_m^{-1})$$

This simple result, of course, could be obtained on the basis of an elementary analysis of deformations.

Consider now the load distribution among the mechanisms along the axis x, which, in accordance with (123) and (124), is characterized by the functions:

$$S_0 = (P_{00}\gamma_0 + P_{01}\gamma_1)GI_0 = B^{-1}(\gamma_1 - A\gamma_0)$$

$$S_1 = (P_{10}\gamma_0 + P_{11}\gamma_1)GI_1 = B^{-1}(\gamma_0 - D\gamma_1)$$

After the substitution of the values A, B, D, we obtain $S_0 = -c_m y(x, \tau)\Pi'/l$; $S_1 = c_m y(x, \tau)/l$; where $y = \Pi'\gamma_0 - \gamma_1$ is the deformation of the distributed mechanisms. Extreme irregularity of load distribution can be judged from the value of the parameter $\xi_y = y(0, \tau)/y(l, \tau)$:

$$\xi_y = (\varsigma_1 + \theta \operatorname{sh}\theta)/(\varsigma_1 \operatorname{ch}\theta + \theta \operatorname{sh}\theta)$$

where $\theta = \sqrt{\zeta_1(1+\zeta_2)}$, $\zeta_1 = c_M(1 + \Pi'^2)\, c_0$, $\zeta_2 = I_1 \Pi'^2 / I_0$.

The analysis shows that already with three mechanisms the error of continual idealization when the deformations and the torques are determined does not exceed 4 percent.

In order to evaluate the influence of clearances we use the criterion K_Δ (see §6). Suppose the function $y(x, t)$ has the form $y_0(x) + y_1(x)\cos\omega t$, and the value of the clearance (reduced to the driven subsystem) is equal to $2\Delta\gamma$. Then

$$K_\Delta(x) = \frac{4\Delta\gamma}{\pi y_1(x)}\sqrt{1 - \left(\frac{y_0(x)}{y_1(x)}\right)^2}$$

We can now identify zones of x in which values of K_Δ are relatively large. It means that in such zones mechanisms should not be located, since their relieving influence on the other mechanisms is rather small and at the same time they may become sources of the vibroimpact excitation for the system.

REFERENCES

1. Artobolevskii, I. I. Theory of machines and mechanisms. Moscow, Nauka, 1975, 639 pages.
2. Babitskii, B. I. Theory of vibroimpact systems. Moscow, Nauka, 1978, 352 pages.
3. Banakh, L. Ia. Study of dynamics of regular and quasi-regular systems using group theory. In: Vibration of complex elastic systems. Moscow, Nauka, 1981, pp. 5–11.
4. Biderman, V. L. Theory of mechanical vibrations. Moscow, Vysshaia shkola, 1980, 408 pages.
5. Bobrov, A. N. and Perchenok, Iu. G. Automated milling machines. Leningrad, Mashinostroyeniye, 1979, 231 pages.
6. Bolotin, V. V. Dynamic stability of elastic systems. Moscow, GITTL, 1956, 600 pages.
7. Veits, V. L. Dynamics of machine aggregates. Leningrad, Mashinostroyeniye, 1969, 370 pages.
8. Veits, V. L., Kochura, A. E., and Martynenko, A. M. Dynamic analysis of machine drives Leningrad, Mashinostroyeniye, 1971, 352 pages.
9. Vibrations in engineering: Handbook. Moscow, Mashinostroyeniye, Volume 1, 1978, 352 pages; Volume 6, 1981, 456 pages.
10. Vul'fson, I. I. Aggregation and decomposing of branched vibratory systems of cyclic mechanisms. Mashinovedeniye, 1980, No. 6, pp. 20–27.
11. Vul'fson, I. I. Dynamic analysis of cyclic mechanisms. Leningrad, Mashinostroyeniye, 1976, 328 pages.
12. Vul'fson I. I. Dynamic analysis of drives and mechanisms constituting multi-contour vibratory systems with variable parameters. Mashinovedeniye, 1978, No.5, pp. 3–8.
13. Vul'fson I. I. Use of hierarchy of dynamic models for vibration analysis of large cyclic systems. In: Machine mechanics. Moscow, Nauka, 1978, No. 53, pp.88–99.
14. Vul'fson, I. I. Use of correction circuits to abate parametric excitation in machine drives. Mezhvuzovskii thematic collection of scientific papers "Vibrotekhnika", 1981, No. 1 (31), pp. 63–70.
15. Vul'fson, I. I. Study of vibrations in a system of interconnected identical cyclic mechanisms. In: Machine mechanics. Moscow, Nauka, 1983, No. 60, pp. 32–39.

16. Vul'fson, I. I. Study of threshold conditions for excitation of combination resonances. Mashinovedeniye, 1980, No. 3, pp. 19–24.
17. Vul'fson, I. I. Conditional dynamic model of multicontour system of mechanisms. Mashinovedeniye, 1982, No. 4, pp. 14–20.
18. Vul'fson, I. I. Method for analysis of dynamic effect from sharp changes of system parameters. In: Nonlinear vibrations and transfer processes in machines. Moscow, Nauka, 1972, pp. 257–268.
19. Vul'fson, I. I. Vibrations of systems with time-dependent parameters. Prikladnaia matematika i mekhanika, 1969, No. 2, Volume 33, pp. 331–337.
20. Vul'fson, I. I. Specifics of frequency spectrum of large complexes of identical cyclic mechanisms. Izv. vuzov. Mashinostroeniye, 1980, No. 10, pp. 34–37.
21. Vul'fson, I. I. Formation of frequency spectra and dynamic coupling of complicated cyclic systems of mechanisms. Mashinovedeniye, 1984, No. 2, pp.3–10.
22. Vul'fson, I. I. Frequency analysis of regular vibratory systems of mechanisms at a programmed translational motion of their common driven link. Izv. vuzov. Mashinostroeniye, 1982, No.12, pp.35–39.
23. Vul'fson, I. I. and Kolovskii, M. Z. Nonlinear problems of machine dynamics. Leningrad, Mashinostroeniye, 1968, 281 pages.
24. Vul'fson, M. N. Study of responses of mechanical systems with distributed spectrum of natural frequencies to a mono-harmonic excitation. In: Vibro-insulation of machines and protection of the human operator. Moscow, Nauka, 1973, pp. 81–86.
25. Gidaspov, I. A. Determining dynamic loads in closed preloaded transmissions. Mashinovedeniye, 1975, No. 6, pp.3–7.
26. Goloskokov, E. G. and Filippov, A. P. Nonstationary vibrations of mechanical systems. Kiev, Naukova dumka, 1966, 336 pages.
27. Vul'fson, I. I., Klimov, V. A., and Krylovidr, K. V. Study of vibroactivity of ring-structure mechanisms considering impacts in clearances. Mashinovedeniye, 1983, No. 1, pp. 18–24.
28. Kobrinskii, A. E., and Kobrinskii, A. A. Vibro-impact systems. Moscow, Nauka, 1973, 591 pages.
29. Kozhevnikov, S. N. Dynamics of machines with elastic elements. Kiev, Izdatel'stvo AN USSR, 1961, 160 pages.
30. Kolovskii, M. Z. Automatic control of vibration protection systems. Moscow, Nauka, 1976, 320 pages.
31. Kolovskii, M. Z. Dynamics of machines. Leningrad, LPI, 1980, 80 pages.
32. Koritysskii, Ia. I. Vibrations in textile machines. Moscow, Mashinostroyeniye, 1973, 320 pages.
33. Levitskii, N. I. Theory of mechanisms and machines. Moscow, Nauka, 1979, 576 pages.
34. Lur'ye, A. I. Analytical mechanics. Moscow, Fizmatgiz, 1961, 824 pages.
35. Mitropol'skii, Iu. A. Problems of asymptotic theory of nonstationary vibrations. Moscow, Nauka, 1964, 432 pages.
36. Moiceyev, N. N. Asymptotic methods of nonlinear mechanics. Moscow, Nauka, 1969, 380 pages.
37. Nonlinear problems of dynamics and strength of machines. Ed.: V. L. Veits, Leningrad, Izdatel'stvo LSU, 1983, 336 pages.
38. Pal'mov, V. A. Vibrations of elasto-plastic bodies. Moscow, Nauka, 1976, 328 pages.
39. Panovko, Ia. G. Basics of applied theory of vibrations and impact. Leningrad, Mashinostroyeniye, 1976, 320 pages.
40. Petruk, A. I. Issues of synthesis of mechanisms of cyclic machines. Kiev, Naukova dumka, 1981, 119 pages.
41. Pervozvanskii, A. A. and Gaitsgori, V. G. Decomposition, aggregation, and approximate optimization. Moscow, Nauka, 1979, 344 pages.
42. Poliudov, A. N. Programmable load-relaxers for cyclic mechanisms. Lvov, Higher School, 1979, 168 pages.

43. Ragul'skis, K. M. Mechanisms on vibratory foundation (Issues of dynamics and stability). Kaunas, AN Lit. SSR, 1963, 232 pages.
44. Sergeyev, V. I. and Iudin, K. M. Study of dynamics of planar mechanisms with clearances. Moscow, Nauka, 1974, 111 pages.
45. Freman, N. and Freman, P.U. WKB-approximation. Moscow, Mir, 1967, 168 pages.
46. Khitrik, V. E. Methods of dynamic optimization of mechanisms for automatic machines. Leningrad, Izdatel'stvo LSU, 1974, 116 pages.
47. Shmidt, G. Parametric vibrations. Moscow, Mir, 1978, 336 pages.
48. Shchepetil'nikov, V. A. Counterbalancing of mechanisms. Moscow, Mashinostroyeniye, 1982, 256 pages.
49. Dresig, H. and Thümel, T. Näherungsweise erfassung des einflusses des gelenkspiels auf die gelerkräfte in schnellaufenden koppelgetrieben-technische mechanik 3, Hf. 1, 1982, pp.33–38.
50. Holzweissig, F. and Dresig, H. Lehrbuch des maschinendynamik. Leipzig, VEB Fachbuchverlag, 1979, 412 pages.
51. Rössler, J. Zur modellierung von schwingungssystemen die periodiskch übersetzende getriebe erhalten-technische mechanik 3, Hf. 1, 1982 pp.39–43.

INDEX